AF361668

Research on Pathogenic Fungi and Mycotoxins in China

Research on Pathogenic Fungi and Mycotoxins in China

Editors

Shihua Wang

Yang Liu

Qi Zhang

MDPI • Basel • Beijing • Wuhan • Barcelona • Belgrade • Manchester • Tokyo • Cluj • Tianjin

Editors

Shihua Wang
Fujian Agriculture and
Forestry University
China

Yang Liu
Foshan University/
South China Food Safety
Research Center
China

Qi Zhang
Chinese Academy of
Agricultural Sciences
China

Editorial Office
MDPI
St. Alban-Anlage 66
4052 Basel, Switzerland

This is a reprint of articles from the Special Issue published online in the open access journal *Toxins* (ISSN 2072-6651) (available at: https://www.mdpi.com/journal/toxins/special_issues/Fungi_Mycotoxins_China).

For citation purposes, cite each article independently as indicated on the article page online and as indicated below:

LastName, A.A.; LastName, B.B.; LastName, C.C. Article Title. *Journal Name* **Year**, *Volume Number*, Page Range.

ISBN 978-3-0365-6269-8 (Hbk)
ISBN 978-3-0365-6270-4 (PDF)

Cover image courtesy of Shihua Wang

Contents

Article

aflN Is Involved in the Biosynthesis of Aflatoxin and Conidiation in *Aspergillus flavus*

Kunzhi Jia [†], Lijuan Yan [†], Yipu Jia, Shuting Xu, Zhaoqi Yan and Shihua Wang *

Key Laboratory of Pathogenic Fungi and Mycotoxins of Fujian Province, Key Laboratory of Biopesticide and Chemical Biology of Education Ministry, School of Life Sciences, Fujian Agriculture and Forestry University, Fuzhou 350002, China; kjia@fafu.edu.cn (K.J.); lyan@fafu.edu.cn (L.Y.); yipujia@fafu.edu.cn (Y.J.); shutingxu@fafu.edu.cn (S.X.); zhaoqiyan@m.fafu.edu.cn (Z.Y.)
* Correspondence: wshmail@m.fafu.edu.cn
† Contributed equally to this work.

Abstract: *Aspergillus flavus* poses a threat to society economy and public health due to aflatoxin production. *aflN* is a gene located in the aflatoxin gene cluster, but the function of AflN is undefined in *Aspergillus flavus*. In this study, *aflN* is knocked out and overexpressed to study the function of AflN. The results indicated that the loss of AflN leads to the defect of aflatoxin biosynthesis. AflN is also found to play a role in conidiation but not hyphal growth and sclerotia development. Moreover, AlfN is related to the response to environmental oxidative stress and intracellular levels of reactive oxygen species. At last, AflN is involved in the pathogenicity of *Aspergillus flavus* to host. These results suggested that AflN played important roles in aflatoxin biosynthesis, conidiation and reactive oxygen species generation in *Aspergillus flavus*, which will be helpful for the understanding of *aflN* function, and will be beneficial to the prevention and control of *Aspergillus flavus* and aflatoxins contamination.

Keywords: *Aspergillus flavus*; AflN; aflatoxin; conidiation

Key Contribution: Our study provides the genetic evidence of *aflN* involvement in the biosynthesis of aflatoxin; conidiation and oxidative stress response in *Aspergillus flavus*; which contributes to the better understanding of *aflN* functions in *A. flavus*.

Citation: Jia, K.; Yan, L.; Jia, Y.; Xu, S.; Yan, Z.; Wang, S. *aflN* Is Involved in the Biosynthesis of Aflatoxin and Conidiation in *Aspergillus flavus*. *Toxins* **2021**, *13*, 831. https://doi.org/10.3390/toxins13110831

Received: 24 October 2021
Accepted: 20 November 2021
Published: 22 November 2021

Publisher's Note: MDPI stays neutral with regard to jurisdictional claims in published maps and institutional affiliations.

1. Introduction

Aspergillus flavus (*A. flavus*) is a notorious pathogenic fungus, which can produce aflatoxins (AFs) and contaminates many crop seeds, leading to the large economic losses [1]. It is worth noting that *A. flavus* is typically found in soil and distributed worldwide due to its strong survival capability [1,2]. Therefore, *A. flavus* poses a threat to society economy and public health. A lasting and deep study on *A. flavus* will help us better understanding and controlling of *A. flavus* and AFs. As the main focus of *A. flavus* study, biosynthesis pathway of AFs is constituted of more than 25 enzymatic reactions [3,4], and these enzyme genes are mainly clustered on chromosome 3 [5]. Many other genes outside the AF gene cluster also have effects on the biosynthesis of AF [6,7]. CreA, which is the master regulator of carbon catabolite repression, was found to regulate the AF biosynthesis and conidia development.

aflN, located at the aflatoxin (AF) gene cluster, is a member of cytochrome P450 family [8]. Previous reports suggested that *aflN* in *A. parasiticus* and *A. nidulans* is involved in the biosynthesis of aflatoxins (AFs), but the detailed functions are still undetermined [3,9]. Genetic disruption of *stcS*, a homolog of *aflN* in *A. nidulans*, led to the accumulation of versicolorin A (VA) and blocked the formation of sterigmatocystin (ST) [9,10], which is an important step for biosynthesis of aflatoxin. Although AflN is suggested to be involved in the biosynthesis of aflatoxin, the direct evidence of AflN involved in AF biosynthesis in *A. flavus* is absent, and the potential other function for AflN is still unclear. In this study, genetic *aflN* mutants were constructed with homology recombination, and the effects of *aflN* on the development and metabolism were then investigated for better understanding of AflN biofunctions in *A. flavus*.

2. Results

2.1. Identification and Analysis of AflN in A. flavus

A. flavus AflN protein was identified from the National Centre for Biotechnology Information (NCBI) database with the sequence ID: XP_002379939.1 (G4B84_005799). Protein sequences from 9 fungi were aligned using MEGA X, and then a phylogenetic tree was constructed. As shown in Figure 1A, *A. flavus* AflN protein shares a 99% similarity with *A. oryzae* AflN, and high similarity with homologues from other 8 fungi, showing that AflN probably plays a similar role in these species. *A. flavus* AlfN has a P450 superfamily domain, highly similar to P450 monooxygenases from other fungi (Figure 1B). To further study the function of AflN, *aflN* knockout (Δ*aflN*), complementary (*aflN-com*) and overexpression (*OE::aflN*) strains have been constructed using homology rearrangement strategy (Figure 1C). The strains' genotyping were verified with PCR method (Figure 1D). Expression levels of *aflN* in various strains were confirmed using qRT-PCR (Figure 1E), showing that *aflN* mutant strains were successfully constructed for further function study.

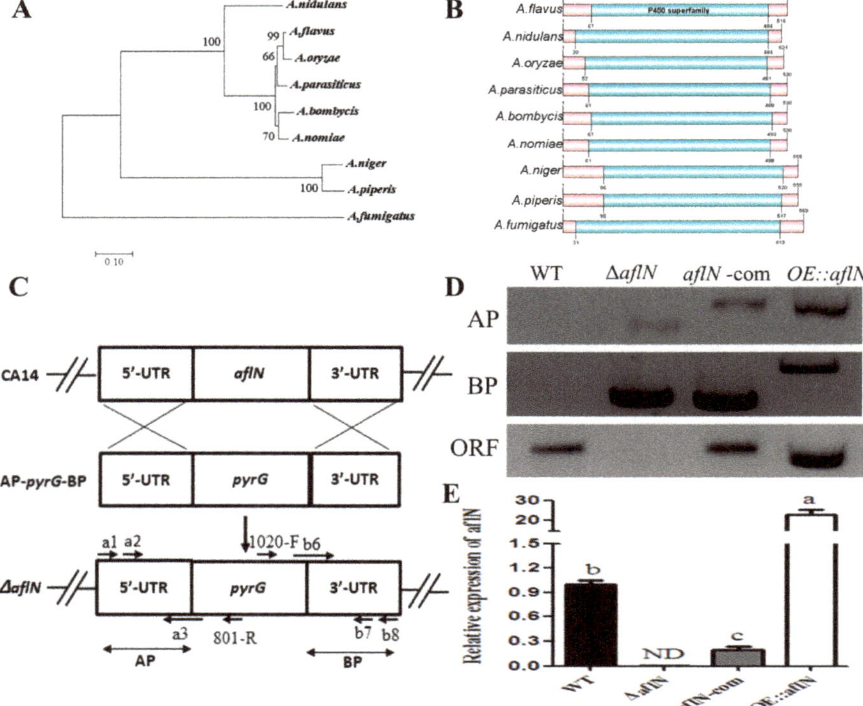

Figure 1. AflN identification and mutant strains construction. (**A**) Phylogenetic tree of AflN homologous proteins from various fungi. The phylogenetic tree was constructed using MEGA X with protein sequences. (**B**) Conserved domain analysis of AflN homologous proteins in different species. (**C**) A typical schematic describing the disruption strategy in this study. UTR represents untranslation region. AP and BP represent A homology arm part and B homology arm part respectively. (**D**) *aflN* mutant strains were verified with PCR. WT means wild type. AP and BP represent A homology arm part and B homology arm part respectively. ORF represents open reading frame of *aflN* gene. (**E**) Expression of *aflN* was examined by qRT-PCR in different *A. flavus* strains. ND means not detected. Different lowercase letters above the bars represent significant difference ($p < 0.05$).

2.2. AflN Plays a Role in the Aflatoxin Biosynthesis in Cytoplasm

To study the effect of AflN on the biosynthesis of AFB1, aflatoxin levels have been assayed in various strains by TLC (thin layer chromatography) method. As shown in Figure 2A, compared with WT and *aflN-com*, AFB1 was barely detected in Δ*aflN* but significantly increased in *OE::aflN*, showing that AflN plays a positive role in the aflatoxin biosynthesis. As known, the location of protein is related with its bio-function. To examine the distribution of AflN, *aflN* fusing with *gfp* was inserted in chromosome using homology rearrange method. As shown in Figure 2B, AflN was distributed in the cytoplasm at 8 h post inoculation of conidia. To make clear whether AflN changed its location at hyphal growth stage, the location of AflN at 24 h post inoculation was also examined. As shown in Figure 2B, AflN was still distributed in the cytoplasm at 24 h. These results indicated that AflN played a critical role in the biosynthesis of AFB1 in cytoplasm.

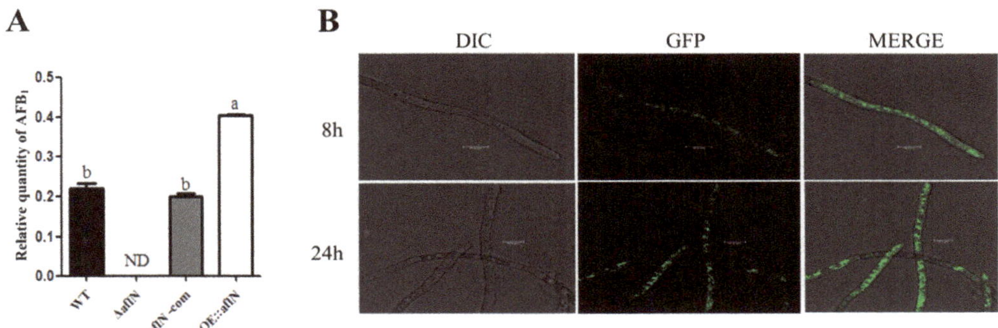

Figure 2. AflN was involved in biosynthesis of aflatoxins. (**A**)The relative quantity of aflatoxins AFB1 in *WT*, Δ*aflN*, *aflN-com* and *OE::aflN*. ND means not detected. Different lowercase letters represent significant difference ($p < 0.05$). (**B**) Location of AflN in *A. flavus* by examing *alfN-gfp* expression with fluorescent microscopy. DIC means differential interference contrast. GFP represents green fluorescent protein.

2.3. AflN Is Involved in Conidiation but Not Hyphal Growth and Sclerotia Development

To study the role of AflN in conidia development, conidiation was studied in various strains. As shown in Figure 3A, the morphology exhibits difference among WT, Δ*aflN*, *aflN-com* and *OE::aflN*. At the same time, the conidia number was significantly decreased in Δ*aflN* but increased in *OE::aflN*, compared with WT and *aflN-com*. This phenomenon was observed both in YGT (yeast extract, glucose and trace elements) and PDA (potato dextrose agar) medium (Figure 3B), which indicated that AflN played an important role in the conidia development. As *brlA* gene plays a critical role in the conidiation [11], *brlA* expression has been examined in this study. As in Figure 3D, *brlA* expression was decreased in Δ*aflN* but increased in *OE::aflN*, compared with WT and *aflN-com*. At the same time, conidiophores in various strains have been observed. As shown in Figure 3E, less conidia and poorly developed conidiophores were observed in Δ*aflN*. These results suggested that the decreased conidiation in Δ*aflN* may be due to the decreased *brlA* expression and poor conidiophores. To investigate whether AflN plays a role in other developmental processes, the hyphal growth was also been examined. The colony diameters were measured, and the result in Figure 3C showed that there is no difference among Δ*aflN*, *aflN-com* and *OE::aflN*. Sclerotia development was also assayed among various strains to determine the role of AflN in this process. As shown in Figure 4A,B, equal levels of sclerotia were observed among Δ*aflN*, *aflN-com* and *OE::aflN* strains. The expression of sclerotia related gene, *nsdD*, exhibited a similar level among various strains (Figure 4C). Thus, AflN has no effect on sclerotia development. All of the above results indicated that AflN has been involved in conidiation but not hyphal growth and sclerotia development in *A. flavus*.

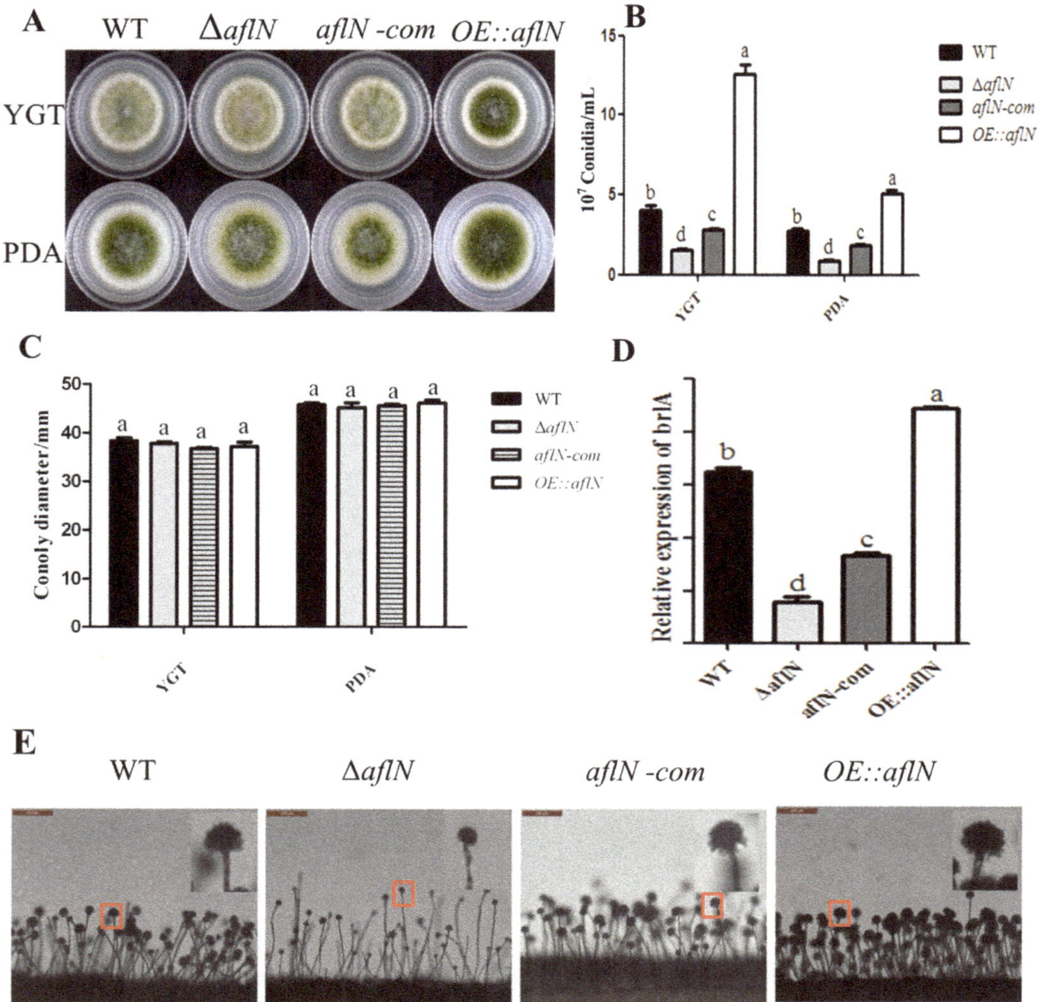

Figure 3. The role of AflN in conidiation and hyphal growth. (**A**) Colony morphology of the WT, Δ*aflN*, *aflN-com* and OE::*aflN* strains grown on YGT and PDA. (**B**) The comparison of conidia number in WT, Δ*aflN*, *aflN-com* and OE::*aflN*. (**C**) The comparison of hyphal growth in WT, Δ*aflN*, *aflN-com* and OE::*aflN*. (**D**) BrlA expression in WT, Δ*aflN*, *aflN-com* and OE::*aflN*. (**E**) Conidiophore morphology of WT, Δ*aflN*, *aflN-com* and OE:: *aflN* strains of *A. flavus* grown on PDA. Different lowercase letters represent significant difference ($p < 0.05$).

2.4. Absence of aflN Leads to Less Response to Oxygen Stress

As cytochrome P450 enzyme, AflN is closely related to the oxygen-reduction system [12]. To study the role of AflN in oxidative stress response, growth of *A. flavus* has been examined in various strains with H_2O_2. As shown in Figure 5A,B, Δ*aflN* strain showed less responsible to 2.5 mM H_2O_2 than WT and *aflN-com*; in contrast, OE::*aflN* showed more sensitive to H_2O_2, indicating that aflN is involved in the response to oxygen stress. Intracellular levels of reactive oxygen species (ROS) reflect the state of oxygen stress to a certain extent [13]. Therefore, ROS levels were further examined using DCFH-DA (6-carboxy-2,7-dichlorodihydrofluorescein diacetate) staining in this study [14]. As shown in

Figure 5C,D, the ROS levels were significantly higher in Δ*aflN* than WT and *aflN-com*. In contrast, the ROS levels were lower in *OE::aflN* strain. To study the reason for high ROS in Δ*aflN*, the expression levels of ROS related genes have been examined. As shown in Figure 5E, compared with WT and *aflN-com*, the expression levels of ROS scavenging enzyme genes, *catalase-like* and *catA*, were significantly decreased in Δ*aflN* but increased in *OE::aflN*. This phenomenon indicated that the ROS scavenging enzymes are related to the increased levels of ROS in Δ*aflN*.

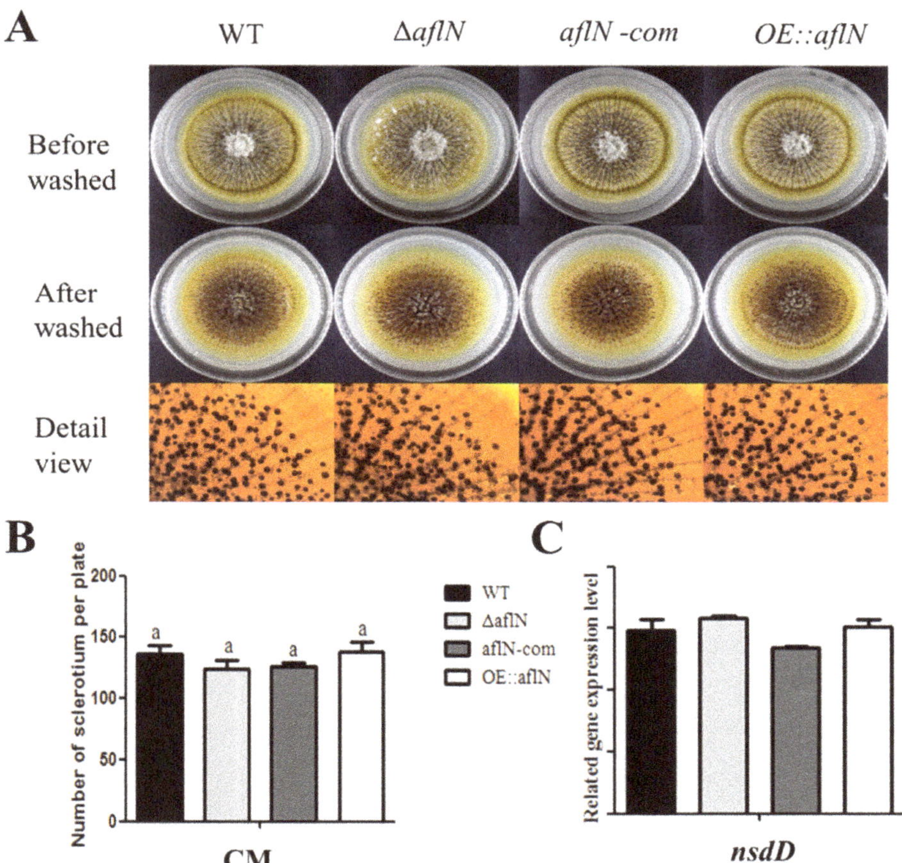

Figure 4. Sclerotia analysis of *A. flavus* among WT, Δ*aflN*, *aflN*-com and *OE::aflN* strains. (**A**) Phenotypes of WT, Δ*aflN*, *aflN*-com and OE::*aflN* strains were determined on CM (complete medium) medium. (**B**) The comparison of sclerotia number among WT, Δ*aflN*, *aflN*-com and *OE::aflN* strains. The lowercase letter "a" represents no significant difference ($p > 0.05$) among various strains. (**C**) Expression levels of *nsdD* involved in sclerotia production by qRT-PCR.

2.5. AflN Is Important for A. flavus Pathogenicity to Crops Seeds

In this study, the pathogenicity of *A. flavus* was assayed by the infection to peanuts and maize [15]. As shown in Figure 6A, the infections of peanuts and maize were observed at the 5th day post-inoculating, and the result indicated that the infection in Δ*aflN* was less than that in WT and *aflN-com*, but the infection in OE::*aflN* was more than that in WT and *aflN-com*. The aflatoxin levels have also been measured by TLC. As shown in Figure 6B,C, compared with WT and *aflN-com* strain, AFB1 in Δ*aflN* infecting peanuts and maize was barely detected, but AFB1 in *OE::aflN* infections significantly increased. At the same time, conidia numbers from the infected peanuts and maize were quantified. As shown in Figure 6B,C, conidia from Δ*aflN* infecting peanuts and maize were significantly less than that in WT and *aflN-com* strain, while conidia from peanuts and maize infected with *OE::aflN* were more than that in WT and *aflN-com* strain. These results indicated that AflN is important for *A. flavus* pathogenicity to crops seeds, and absence of AflN leads to the impairment of pathogenicity.

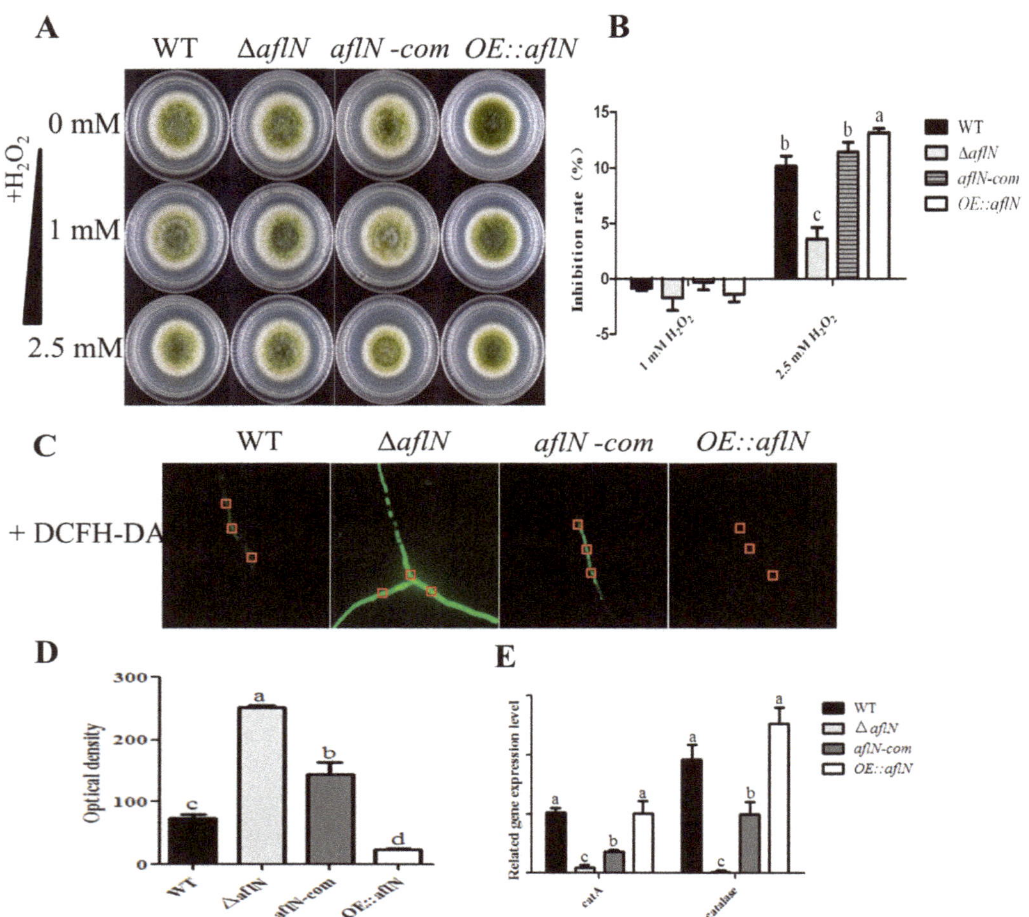

Figure 5. The role of *aflN* in oxidative stress. (**A**) Strains were treated with 1 mM H₂O and 2.5 mM H₂O₂ stress. (**B**) The comparison of the growth inhibition rate among WT, Δ*aflN*, *aflN*-com and *OE::aflN* strains treated with H₂O₂. (**C**) ROS signal was indicated with DCFH-DA staining in various strains. (**D**) Quantitative analysis of ROS signals in various strains. (**E**) Expression levels of oxidation-related enzyme genes (*catalase like* and *catA*) in various strains. Different lowercase letters represent significant difference (*p* < 0.05).

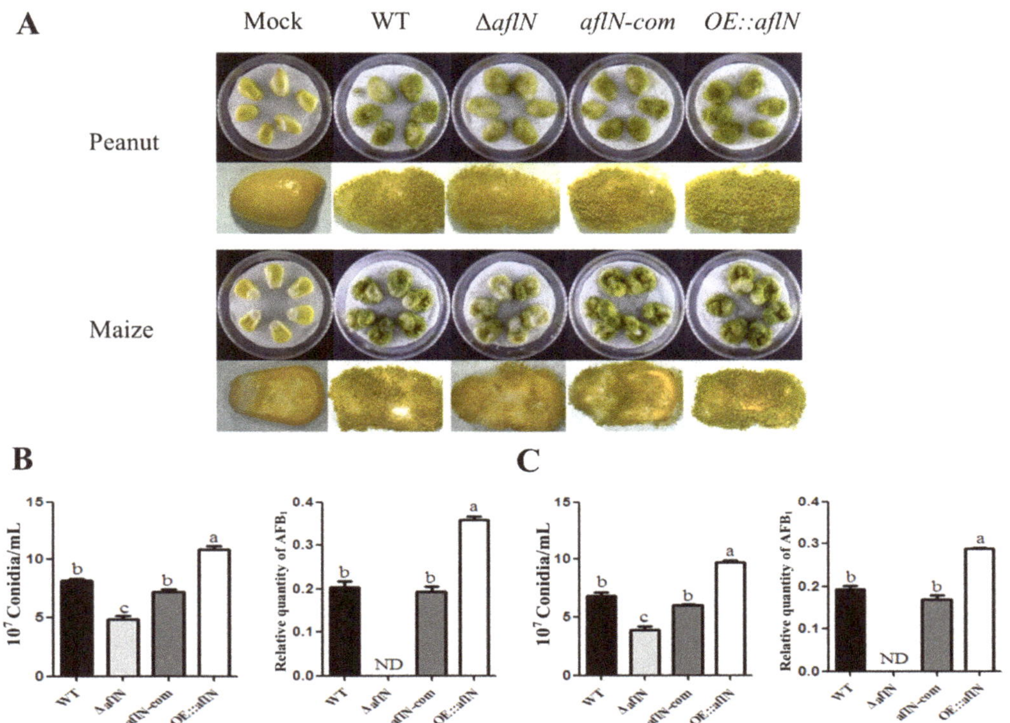

Figure 6. The effect of *aflN* gene on the pathogenicity of *A. flavus*. (**A**) Infection of peanuts and maize with various strains. Mock represents a control without any AF strain's infection. (**B**) The number of conidia and the relative quantity of AFB1 in infected peanuts. (**C**) The number of conidia and the relative quantity AFB1 in infected maize. Different lowercase letters represent significant difference ($p < 0.05$).

3. Discussion

Genetic disruption of *stcS* was found to block the conversion of VA to ST in *A. nidulans* [9]. Then, *StcS* was suggested to be a homology of *verA* (*aflN*) based on the high similarity of their protein sequences [3,10]. VerA (AflN) in *A. parasiticus* is proposed to be involved in the conversion of VA to ST, one of undefined steps before ST formation. In this study, AflN was identified in *A. flavus* by alignment of protein sequence with AflN in other fungi. Phylogenetic tree analysis of AflN indicated that *A.flavus* AflN showed high homology with AflN in *A.oryzae* and *A.parasiticus* (Figure 1A,B), suggesting a potential function in the conversion of VA to ST, a step of AF biosynthesis. To study the function of AflN in *A.flavus*, Δ*aflN*, *aflN-com* and *OE::aflN* strains were successfully constructed with homology recombinant method. These mutant strains were confirmed with PCR, and *aflN* expression were also quantified using qRT-PCR in various strains, which indicated that Δ*aflN*, *aflN-com* and OE::*aflN* strains were successfully constructed (Figure 1D,E).

For *A. flavus*, AF biosynthesis is one of the main characteristics, which were under intense investigation. During the biosynthesis, the conversion of VA to ST is one of complicated and undefined steps, in which AflY, AflX, and AflM are all involved [16–18]. In *A. nidulans*, *stcS*, the homologues of *aflN*, was suggested to be involved in the conversion of VA to ST [9]. In this study, AflN was found to play a role in AF biosynthesis (Figure 2A), consistent with the function of the homology StcS in *A. nidulans*. As the homology of StcS, the absence of AflN possibly causes the defect of the formation of ST, leading to the undetected level of AF, which suggests a direct role of AflN in AF biosynthesis in *A.flavus*.

AflN distribution is important for its function. As shown in Figure 2B, our result indicated that AflN is located at the cytoplasm. The location of AflN is consistent and stationary at 8 h and 24 h, suggesting the function of AflN will be concentrated at cytoplasm. As a member of cytoplasmic P450 superfamily, AflN possibly plays its role at the cytoplasm [19,20]. Consistently, AflN was involved in aflatoxin biosynthesis (Figure 2), which is mainly taking place in cytoplasm [7,21]. Moreover, aflatoxisomes (aflatoxin-synthesizing vesicles) were proved to be units capable of synthesizing aflatoxins including AFB1 [22,23]. Ver-1 has been found in cytoplasm, vesicles and vacuoles in *A.parasiticus* [22]. Thus, as a middle enzyme of AFB1 biosynthesis, it is possible that AflN is also located at aflatoxisomes for aflatoxins synthesis in *A. flavus*. Interestingly, AflN tends to aggregate at undefined places in hypal cells (Figure 2B). To assume that these undefined places partly overlap with aflatoxisomes is reasonable, although it needs further confirmation. Therefore, our results suggested that AflN located in cytoplasm plays its role in the AF biosynthesis in *A. flavus*, beneficial to the further function study of AflN. Moreover, our results provide the genetic evidence for the role of *aflN* in the AF biosynthesis in *A. flavus*.

As known, conidia are the important form for the spread of *A. flavus* [24]. Our study also found that AflN is involved in the conidiation. The poorly developed conidiophores and decreased *brlA* expression were also observed in Δ*aflN*. Considering that conidia development is mainly regulated by *brlA* [11], this impairment of conidiation is possibly due to the decreased expression of *brlA* in Δ*aflN*. Moreover, BrlA is necessary and sufficient for conidiophore development [11]. Therefore, decreased *brlA* expression plays a crucial role in the abnormal development of conidia in Δ*aflN*. However, it is impossible to ignore the accumulation of versicolorin A, the substrate of AflN [8], and the absence of AFB1 (Figure 2A) in Δ*aflN*. Although no evidence suggests that versicolorin A inhibit conidiophore, the possibility cannot be excluded. At the same time, reduced AFB1 is always associated with poor conidiophores and impaired conidiogenesis [6,25]. Although the detailed mechanism may be different, AFB1 seems to be a beneficial signal for conidiogenesis. The roles of AflN in hyphal growth and sclerotia were also investigated in this study. Our results demonstrated that AflN has no effect on the hyphal growth and sclerotial development in *A.flavus* (Figure 4). All together, these results indicated that AflN plays an important role in conidiation but not in hyphal growth and sclerotia development.

During the vegetative growth, *A. flavus* undergoes and responds to various environmental stresses, which may be closely related to AF biosynthesis [26]. In this study, the inhibiting effect of oxidative stress on *A. flavus* was decreased in Δ*aflN* (Figure 5A,B). At the same time, the increased ROS has been observed in Δ*aflN* (Figure 5C,D), which poses *A. flavus* the state of anti-ROS for the balance of oxygen-reducing system. The increased ROS may be partly due to the decreased ROS scavenging enzymes, catA and catalase-like in Δ*aflN* (Figure 5E), but this does not exclude other reasons. It is worth noting that aflatoxins biosynthesis also has close relationship with ROS. It was proposed that aflatoxin biosynthesis is part of the cellular response to oxidative stress [27,28]. High level of ROS is able to trigger aflatoxin biosynthesis associated with increased expression of aflatoxin cluster genes [28]. The upregulation of gene expression is regulated by transcription factors including AtfB, SrrA, AP-1 and MsnA, which are triggered by the increased ROS [28]. *fas-1*, *omtA* and *ver-1* in AF gene cluster were proved to respond to the regulation factors [28]. Interesting, upregulation of *aflP* (*omtA*) and *aflM* (*ver-1*) have been observed in Δ*aflN* (unpulished data), although related transcriptional factors have not been examined. How the absence of AflN affects the expression of ROS scavenging enzymes is unknown and this demands further study. All of these results indicated that *aflN* is involved in anti-oxygen stress in *A. flavus*.

The AFB1 contamination of food is usually concerned with prevention and control of *A. flavus* in agriculture. In this study, AFB1 were undetected in peanuts and maize infected with Δ*aflN* (Figure 6), significantly decreasing the virulence and pathogenicity of *A.flavus*. At the same time, conidia number was decreased in Δ*aflN* infection, which will restrict the dispersion of *A.flavus* [29], compared with WT. The impaired pathogenicity in Δ*aflN* is possibly due to the role of AflN in AFB1 biosynthesis and conidiation. These results also suggested that study of *aflN* function has the potential value in the prevention and control of *A. flavus* and aflatoxin contamination.

4. Conclusions

In conclusion, *aflN* is involved in the AF biosynthesis, oxidative stress response, conidiation and pathogenicity, but not in growth and sclerotia formation. These findings contribute to the better understanding of *aflN* functions in *A. flavus*, which potentially helps to provide a reference for scientific prevention and control of *A. flavus*.

5. Materials and Methods

5.1. Mycelia Growth, Conidiation and Sclerotia Production

A. flavus strains were listed in Table 1. For mycelia growth assays, various strains were cultured on solid YES (yeast extract-sucrose) and PDA (potato dextrose agar) at 37 °C for indicated time. Then colony diameters were observed and measured [30]. For conidia assay, YGT (yeast extract, glucose and trace elements) and PDA were used [6]. For sclerotial production analysis, strains were grown on solid CM (complete medium) at 37 °C for 7 days in the dark [6].

Table 1. *A. flavus* strains used in this study.

Strain	Genotype Description	Reference
CA14 PTS	Δ*ku70*, Δ*niaD*, Δ*pyrG*	[31]
wild-type	Δ*ku70*, Δ*niaD*, Δ*pyrG*:: *pyrG*	This study
Δ*aflN*	Δ*ku70*, Δ*niaD*, Δ*pyrG*, Δ*aflN*:: *pyrG*	This study
aflN-com	Δ*ku70*, Δ*niaD*, Δ*pyrG*::*aflN*, *pyrG*	This study
OE::aflN	Δ*ku70*, Δ*niaD*, Δ*pyrG*::Gpda(p)-*aflN*, *pyrG*	This study
aflN-gfp	Δ*ku70*, Δ*niaD*, Δ*pyrG*:: *aflN-gfp*, *pyrG*	This study

5.2. Construction of Gene Mutants

The *aflN* deletion (Δ*aflN*), complementary (*aflN-com*) and overexpression (*OE::aflN*) strains were constructed with the methods previously described [6,25,31]. Briefly, the upstream (AP, A homology arm part) and downstream (BP, B homology arm part) sequences of *aflN* gene, as well as the *pyrG* gene from *A. fumigatus*, were fused into an interruption fragment (AP-*pyrG*-BP) using fusion PCR strategy. The AP fragment was amplified using primers *aflN-a1* and *aflN-a3*, while the BP fragment was amplified using primers *aflN-b6* and *aflN-b8*. *pyrG* gene was amplified using primers *pyrG-F* and *pyrG-R*. *aflN-b2* and *aflN-b7* are nesting primers for further amplifing AP-*pyrG*-BP fragment, which is from the overlap extension PCR of AP, BP, and *pyrG*. The fused fragment (AP-*pyrG*-BP) was then transformed into *A. flavus* CA14 PTs protoplasts to construct *aflN* knockout strain (Δ*aflN*). The Δ*aflN* strain was primarily confirmed with PCR method. AP, BP and ORF (open reading frame for aflN) fragments were amplified using primers *aflN-a1/P801-R*, *P1020-F/aflN-b8*, and *aflN-F4/aflN-F5*, respectively, to validate if the insertion takes place at the designed site. For the construction of complementary strain (*aflN-com*), two steps of homology recombination were used. First, the *pyrG* gene in Δ*aflN* was replaced with *aflN* gene fragment, which is confirmed by PCR method. Second, *pyrG* gene was inserted into the site between *aflN* ORF and 3′UTR of the intermediate strain for the selection of *aflN*-com strain. For the construction of overexpression strain (*OE::aflN*), *aflN* gene with *gpdA* promoter from *A. nidulans* was amplified, and the fused fragment (AP-*pyrG-gpdA*-BP) was transformed into A. flavus CA14 PTs protoplasts to construct *OE::aflN*. All mutant strains were confirmed by PCR, sequencing, and qRT-PCR. Primers used for *aflN* mutant strains construction and verification are listed in Table 2. For the localization of AflN, *aflN-gfp* fusion cassette was constructed as previously described [32].

Table 2. Primers for strains construction.

Primer Name	Sequence(5′→3′)	Application
aflN-a1	AGGTATTCAGATATTTCGGTCTC	
aflN-a3	GGGTGAAGAGCATTGTTTGAGGCG GTCATGTCCCTAGTTCGT	
aflN-b6	GCATCAGTGCCTCCTCTCAGAC ATTTGTAAGAATGTCGTGCCT	For Δ*aflN* construction and verification
aflN-b8	GTCGCGGGAGGAAATGA	
aflN-a2	ATCCTGACCAGCTCTAA	
aflN-b7	CCTTTCCAAACCCTAC	
aflN-F4	TTCCTGACGGCGTTCTA	
aflN-R5	CACGATGCCCATTGACTT	
com-aflN-F	GCAGCCACCCAAATACAAAAGT	
com-aflN-R	GGGTGAAGAGCATTGTTTGAGGCC ATGACCCTCACTAAAACTACCCT	
P1020-F	ATCGGCAATACCGTCCAGAAGC	For *aflN-com* construction and verification
P801-R	CAGGAGTTCTCGGGTTGTCG	
pyrG-F	GCCTCAAACAATGCTCTTCACCC	
pyrG-R	GTCTGAGAGGAGGCACTGATGC	
qPCR-*aflN*-F	TCCTCTCGAGTCGCTCACCAC	
qPCR-*aflN*-R	ACCATAGTACCAACGGCCTAA	
gpdA-F	GCATCAGTGCCTCCTCTCAGACG AGGACTGCAATCGCCATGAGGTTT	
gpdA-R	CAAGCTGCGATGAAGTGGGAAAG	
aflN-OE-AF	AGTGGTTGAACAGATCAAGGC	
aflN-OE-AR	GAAGAGCATTGTTTGAGGCC TATACATCGTCAGCTTCAGGA	For *OE::aflN* construction and verification
aflN-OE-BF	CAAAGAGCAAACCTTC CTATGCCAGAGTTCAAGCT	
aflN-OE-BR	TGAGAACAGGAGATAGACAGC	
aflN-OE-MF	CGCTTGAGCAGACATCACA ATGTACCTTTCGCTCCTCAT	
aflN-OE-MR	GAACTCTGGCATAGGAA GGTTTGCTCTTTGCAGC	
aflN-OE-P2	GAGGCCTATCGCCATATGCG	
aflN-OE-P7	CACTCATCGTATGCTGGCG	

5.3. The Levels of Intracellular ROS

The intracellular ROS level was measured by using a ROS assay kit (Beyotime Institute of Biotechnology, Shanghai, China) with the protocol manual. In brief, the harvested mycelia were incubated with PBS (phosphate buffered saline) containing 10 μM DCFH-DA (6-carboxy-2,7-dichlorodihydrofluorescein diacetate) for 30 min. The fluorescence signal of ROS level was acquired by confocal microscope.

5.4. Stress Analysis

Various strains in this study were pointed onto solid PDA supplemented with oxidative stress-inducing reagents (1 mM or 2.5 mM H_2O_2). Then the plates were incubated at 37 °C in the dark for 4 days before the calculation of relative inhibition.

5.5. AF Analysis

AF was extracted from liquid YES medium using chloroform [33]. TLC was used to analyze the level of AF biosynthesis, as previously reported [30]. In brief, 5 μL of AF suspension was loaded into a silica gel plate and separated by chromatographic solution including acetone: chloroform (1:9, *v/v*). Silica gel plates were examined with Gel Doc XR+(Bio-Rad) and captured at 312 nm wavelength. The AFB1 signal was analyzed and quantified with Image J.

5.6. The Infection of Peanuts and Maize Seeds

To assess the pathogenicity of *A. flavus*, peanut and maize seeds, which were kept in our lab and used for the infection, were washed with sodium hypochlorite, ethanol and sterile water. The washed seeds were then infected by immersing in a 10^7 spores suspension for 30 min. Then, the seeds were placed in culture dishes lined with moist sterile filter paper for 5 days. The infected seeds were pictured and collected for spores count and AFB1 quantification.

5.7. qRT-PCR

Total RNA was prepared from the mycelia of *A. flavus* with the use of RNA extraction kit (TIANMO BIOTECH, Beijing, China). SYBR Green Supermix (Takara) was used for the qRT-PCR reaction with the PikoReal 96 Real-time PCR system. The $2^{-\Delta\Delta CT}$ method was used to quantify the expression levels of the indicated genes [34]. PCR primers for qRT-PCR were shown in Table 3.

Table 3. Primers for qRT-PCR.

Primers Name	Sequence	Application
brlA-F	GCCTCCAGCGTCAACCTTC	*brlA*
brlA-R	TCTCTTCAAATGCTCTTGCCTC	
nsdD-F	GGACTTGCGGGTCGTGCTA	*nsdD*
nsdD-R	AGAACGCTGGGTCTGGTGC	
Catalase-F	TCGAACAATTCCGTGGTATG	*Catalase like*
Catalase-R	AGCTGGTCGCTCCCGATGGA	
CatA-F	CGCCATCATTATCGGCGACGGA	*CatA*
CatA-R	TGAGGCTTTCGACGTGCGGAC	
β-tublin-F	TTGAGCCCTACAACGCCACT	*β-tublin*
β-tublin-R	TGGTTCAGGTCACCGTAAGAGG	
Actin-F	ACGGTGTCGTCACAAACTGG	*Actin*
Actin-R	CGGTTGGACTTAGGGTTGATAG	

5.8. Statistical Analysis

Data are presented as means $\pm$ SD. All statistics were performed using Student's *t* test with SPSS 11.5 software. A $p < 0.05$ was considered statistically significant.

Author Contributions: K.J. and S.W. designed the experiments and wrote the manuscript. L.Y., Y.J., S.X. and Z.Y. performed all the experiments. All authors have read and agreed to the published version of the manuscript.

Funding: This work was funded by the National Natural Science Foundation of China (31172297), education and research projects for young teachers in Fujian Provincial Education Hall (JT180137), Key Laboratory of Ministry of Education for Genetics, Breeding and Multiple Utilization of Crops, College of Agriculture, Fujian Agriculture and Forestry University (GBMUC-2019-003), and the science & technology innovation fund of Fujian Agriculture and Forestry University (CXZX2017512, KFA17583A).

Institutional Review Board Statement: Not applicable.

Informed Consent Statement: Not applicable.

Conflicts of Interest: The authors declare no conflict of interest.

References

1. Dhakal, A.; Sbar, E. Aflatoxin Toxicity. In *StatPearls [Internet]*; StatPearls Publishing: Treasure Island, FL, USA, 2021.
2. Kumar, P.; Mahato, D.K.; Kamle, M.; Mohanta, T.K.; Kang, S.G. Aflatoxins: A Global Concern for Food Safety, Human Health and Their Management. *Front. Microbiol.* **2016**, *7*, 2170. [CrossRef] [PubMed]
3. Zeng, H.; Cai, J.; Hatabayashi, H.; Nakagawa, H.; Nakajima, H.; Yabe, K. verA Gene is Involved in the Step to Make the Xanthone Structure of Demethylsterigmatocystin in Aflatoxin Biosynthesis. *Int. J. Mol. Sci.* **2020**, *21*, 6389. [CrossRef] [PubMed]

4. Yabe, K.; Nakajima, H. Enzyme reactions and genes in aflatoxin biosynthesis. *Appl. Microbiol. Biotechnol.* **2004**, *64*, 745–755. [CrossRef] [PubMed]
5. Bhatnagar, D.; Cary, J.W.; Ehrlich, K.; Yu, J.; Cleveland, T.E. Understanding the genetics of regulation of aflatoxin production and *Aspergillus flavus* development. *Mycopathologia* **2006**, *162*, 155–166. [CrossRef] [PubMed]
6. Fasoyin, O.E.; Wang, B.; Qiu, M.; Han, X.; Chung, K.R.; Wang, S. Carbon catabolite repression gene creA regulates morphology, aflatoxin biosynthesis and virulence in Aspergillus flavus. *Fungal Genet. Biol.* **2018**, *115*, 41–51. [CrossRef]
7. Yang, M.; Zhu, Z.; Bai, Y.; Zhuang, Z.; Ge, F.; Li, M.; Wang, S. A novel phosphoinositide kinase Fab1 regulates biosynthesis of pathogenic aflatoxin in Aspergillus flavus. *Virulence* **2021**, *12*, 96–113. [CrossRef]
8. Henry, K.M.; Townsend, C.A. Synthesis and fate of o-carboxybenzophenones in the biosynthesis of aflatoxin. *J. Am. Chem. Soc.* **2005**, *127*, 3300–3309. [CrossRef]
9. Keller, N.P.; Segner, S.; Bhatnagar, D.; Adams, T.H. stcS, a putative P-450 monooxygenase, is required for the conversion of versicolorin A to sterigmatocystin in Aspergillus nidulans. *Appl. Environ. Microbiol.* **1995**, *61*, 3628–3632. [CrossRef]
10. Yu, J.; Chang, P.K.; Ehrlich, K.C.; Cary, J.W.; Bhatnagar, D.; Cleveland, T.E.; Payne, G.A.; Linz, J.E.; Woloshuk, C.P.; Bennett, J.W. Clustered pathway genes in aflatoxin biosynthesis. *Appl. Environ. Microbiol.* **2004**, *70*, 1253–1262. [CrossRef]
11. Adams, T.H.; Boylan, M.T.; Timberlake, W.E. brlA is necessary and sufficient to direct conidiophore development in Aspergillus nidulans. *Cell* **1988**, *54*, 353–362. [CrossRef]
12. Hrycay, E.G.; Bandiera, S.M. Involvement of Cytochrome P450 in Reactive Oxygen Species Formation and Cancer. *Adv. Pharmacol.* **2015**, *74*, 35–84. [CrossRef]
13. Lennicke, C.; Cocheme, H.M. Redox signalling and ageing: Insights from Drosophila. *Biochem. Soc. Trans.* **2020**, *48*, 367–377. [CrossRef]
14. Jia, K.Z.; Jin, S.L.; Yao, C.; Rong, R.; Wang, C.; Du, P.; Jiang, W.H.; Huang, X.F.; Hu, Q.G.; Miao, D.S.; et al. Absence of PTHrP nuclear localization and C-terminus sequences leads to abnormal development of T cells. *Biochimie* **2017**, *138*, 13–19. [CrossRef]
15. Yang, K.; Qin, Q.; Liu, Y.; Zhang, L.; Liang, L.; Lan, H.; Chen, C.; You, Y.; Zhang, F.; Wang, S. Adenylate Cyclase AcyA Regulates Development, Aflatoxin Biosynthesis and Fungal Virulence in Aspergillus flavus. *Front. Cell. Infect. Microbiol.* **2016**, *6*, 190. [CrossRef]
16. Ehrlich, K.C.; Montalbano, B.; Boue, S.M.; Bhatnagar, D. An aflatoxin biosynthesis cluster gene encodes a novel oxidase required for conversion of versicolorin a to sterigmatocystin. *Appl. Environ. Microbiol.* **2005**, *71*, 8963–8965. [CrossRef]
17. Skory, C.D.; Chang, P.K.; Cary, J.; Linz, J.E. Isolation.n and characterization of a gene from Aspergillus parasiticus associated with the conversion of versicolorin A to sterigmatocystin in aflatoxin biosynthesis. *Appl. Environ. Microbiol.* **1992**, *58*, 3527–3537. [CrossRef]
18. Cary, J.W.; Ehrlich, K.C.; Bland, J.M.; Montalbano, B.G. The aflatoxin biosynthesis cluster gene, aflX, encodes an oxidoreductase involved in conversion of versicolorin A to demethylsterigmatocystin. *Appl. Environ. Microbiol.* **2006**, *72*, 1096–1101. [CrossRef]
19. Guengerich, F.P.; Waterman, M.R.; Egli, M. Recent Structural Insights into Cytochrome P450 Function. *Trends Pharmacol. Sci.* **2016**, *37*, 625–640. [CrossRef]
20. Jia, K.; Li, L.; Liu, Z.; Hartog, M.; Kluetzman, K.; Zhang, Q.Y.; Ding, X. Generation and characterization of a novel CYP2A13–transgenic mouse model. *Drug Metab. Dispos.* **2014**, *42*, 1341–1348. [CrossRef]
21. Reverberi, M.; Punelli, M.; Smith, C.A.; Zjalic, S.; Scarpari, M.; Scala, V.; Cardinali, G.; Aspite, N.; Pinzari, F.; Payne, G.A.; et al. How peroxisomes affect aflatoxin biosynthesis in Aspergillus flavus. *PLoS ONE* **2012**, *7*, e48097. [CrossRef]
22. Chanda, A.; Roze, L.V.; Kang, S.; Artymovich, K.A.; Hicks, G.R.; Raikhel, N.V.; Calvo, A.M.; Linz, J.E. A key role for vesicles in fungal secondary metabolism. *Proc. Natl. Acad. Sci. USA* **2009**, *106*, 19533–19538. [CrossRef] [PubMed]
23. Roze, L.V.; Chanda, A.; Linz, J.E. Compartmentalization and molecular traffic in secondary metabolism: A new understanding of established cellular processes. *Fungal Genet. Biol.* **2011**, *48*, 35–48. [CrossRef] [PubMed]
24. Adams, T.H.; Wieser, J.K.; Yu, J.H. Asexual sporulation in Aspergillus nidulans. *Microbiol. Mol. Biol. Rev.* **1998**, *62*, 35–54. [CrossRef] [PubMed]
25. Liang, L.; Liu, Y.; Yang, K.; Lin, G.; Xu, Z.; Lan, H.; Wang, X.; Wang, S. The Putative Histone Methyltransferase DOT1 Regulates Aflatoxin and Pathogenicity Attributes in Aspergillus flavus. *Toxins* **2017**, *9*, 232. [CrossRef]
26. Fountain, J.C.; Scully, B.T.; Ni, X.; Kemerait, R.C.; Lee, R.D.; Chen, Z.Y.; Guo, B. Environmental influences on maize-Aspergillus flavus interactions and aflatoxin production. *Front. Microbiol.* **2014**, *5*, 40. [CrossRef]
27. Roze, L.V.; Hong, S.Y.; Linz, J.E. Aflatoxin biosynthesis: Current frontiers. *Annu. Rev. Food Sci. Technol.* **2013**, *4*, 293–311. [CrossRef]
28. Hong, S.Y.; Roze, L.V.; Wee, J.; Linz, J.E. Evidence that a transcription factor regulatory network coordinates oxidative stress response and secondary metabolism in aspergilli. *Microbiologyopen* **2013**, *2*, 144–160. [CrossRef]
29. Dijksterhuis, J. Fungal spores: Highly variable and stress-resistant vehicles for distribution and spoilage. *Food Microbiol.* **2019**, *81*, 2–11. [CrossRef]
30. Yang, K.; Liang, L.; Ran, F.; Liu, Y.; Li, Z.; Lan, H.; Gao, P.; Zhuang, Z.; Zhang, F.; Nie, X.; et al. The DmtA methyltransferase contributes to Aspergillus flavus conidiation, sclerotial production, aflatoxin biosynthesis and virulence. *Sci. Rep.* **2016**, *6*, 23259. [CrossRef]
31. Chang, P.K.; Scharfenstein, L.L.; Wei, Q.; Bhatnagar, D. Development and refinement of a high-efficiency gene-targeting system for Aspergillus flavus. *J. Microbiol. Methods* **2010**, *81*, 240–246. [CrossRef]

32. Zheng, W.; Zhao, X.; Xie, Q.; Huang, Q.; Zhang, C.; Zhai, H.; Xu, L.; Lu, G.; Shim, W.B.; Wang, Z. A conserved homeobox transcription factor Htf1 is required for phialide development and conidiogenesis in Fusarium species. *PLoS ONE* **2012**, *7*, e45432. [CrossRef]
33. Zhu, Z.; Yang, M.; Bai, Y.; Ge, F.; Wang, S. Antioxidant-related catalase CTA1 regulates development, aflatoxin biosynthesis, and virulence in pathogenic fungus Aspergillus flavus. *Environ. Microbiol.* **2020**, *22*, 2792–2810. [CrossRef]
34. Livak, K.J.; Schmittgen, T.D. Analysis of relative gene expression data using real-time quantitative PCR and the 2(-Delta Delta C(T)) Method. *Methods* **2001**, *25*, 402–408. [CrossRef]

Article

Development of Generic Immuno-Magnetic Bead-Based Enzyme-Linked Immunoassay for Ustiloxins in Rice Coupled with Enrichment

Yi Huang [1,2], Xiaoqian Tang [2], Lu Zheng [1], Junbin Huang [1], Qi Zhang [2,*] and Hao Liu [1,*]

[1] The Key Lab of Plant Pathology of Hubei Province, Huazhong Agricultural University, Wuhan 430070, China; hzauhuangyi@163.com (Y.H.); luzheng@mail.hzau.edu.cn (L.Z.); junbinhuang@mail.hzau.edu.cn (J.H.)

[2] Oil Crops Research Institute, Chinese Academy of Agricultural Sciences, Wuhan 430062, China; wtxqtutu@163.com

* Correspondence: zhangqi01@caas.cn (Q.Z.); HL210@mail.hzau.edu.cn (H.L.)

Abstract: Ustiloxins are a group of mycotoxins produced by rice false smut pathogen. Previous studies have shown that the false smut balls contain six types of ustiloxins, and these toxins are toxic to living organisms. Thus, immunoassay for on-site monitoring of ustiloxins in rice is urgently required. The current immunoassays are only for detecting single ustiloxin, and they cannot meet the demand for synchronous and rapid detection of the group toxins. Therefore, this study designed and synthesized a generic antigen with ustiloxin G as material based on the common structure of the mycotoxins. Ustiloxin G was conjugated to two carrier proteins including bovine serum albumin (BSA) and ovalbvmin (OVA) by carbon diimide method. The mice were immunized with ustiloxin-G-BSA to generate the antibody serum, which was further purified to obtain the generic antibody against ustiloxins. The conjugated ustiloxin G-OVA and generic antibodies were used for establishing the enzyme-linked immunosorbent assay (ELISA) for ustiloxin detection and optimizing experiment conditions. The characterization of the antibody showed that the semi-inhibitory concentrations (IC_{50}) of ustiloxin A, B, and G were 0.53, 0.34, and 0.06 μg/mL, respectively, and that their corresponding cross-reactivities were 11.9%, 18.4%, and 100%, respectively. To increase ELISA detection efficiency, generic antibody was combined with magnetic beads to obtain sensitive and class-specific immune-magnetic beads. Based on these immuno-magnetic beads, a high-efficiency enzyme-linked immunoassay method was developed for ustiloxin detection, whose sensitivity to ustiloxin A, B, and G was improved to 0.15 μg/mL, 0.14 μg/mL, and 0.04 μg/mL, respectively. The method accuracy was evaluated by spiking ustiloxin G as standard, and the spiked samples were tested by the immune-magnetic bead-based ELISA. The result showed the ustiloxin G recoveries ranged from 101.9% to 116.4% and were accepted by a standard HPLC method, indicating that our developed method would be promising for on-site monitoring of ustiloxins in rice.

Keywords: rice false smut; ustiloxins; generic antigen; immuno-magnetic beads; enzyme-linked immunity

Key Contribution: We provided a new, sensitive and generic immunoassay which can be used for on-site monitoring ustiloxins in rice.

Citation: Huang, Y.; Tang, X.; Zheng, L.; Huang, J.; Zhang, Q.; Liu, H. Development of Generic Immuno-Magnetic Bead-Based Enzyme-Linked Immunoassay for Ustiloxins in Rice Coupled with Enrichment. *Toxins* **2021**, 13, 907. https://doi.org/10.3390/toxins 13120907

Received: 1 November 2021
Accepted: 15 December 2021
Published: 17 December 2021

Publisher's Note: MDPI stays neutral with regard to jurisdictional claims in published maps and institutional affiliations.

1. Introduction

Rice false smut (RFS) has become an emerging and economically important fungal disease in most rice-producing countries recently [1–6]. Since the 1970s, the harm of RFS has been widely reported in the major planting areas in China. After the 1980s, with the change of farming methods and the large-scale promotion of hybrid rice varieties, RFS has exhibited an increasing trend year by year, and currently, it harms nearly one third of the rice cultivation area [7–9]. The damage of RFS has been constantly investigated. In addition to causing the obvious yield loss, it can also produce large amounts of toxins to pollute grains [10–13]. Of all the known toxins produced by RFS pathogen, ustiloxins are the most widely concerned.

Ustiloxins are a group of cyclopeptides containing a 13-membered cyclic core structure with an ether linkage. To date, six types of ustiloxins have been identified and named as ustiloxin A, B, C, D, F, and G, respectively [11,14–16]. Among them, ustiloxin A and B are the main species, accounting for about 90% of the total content of ustiloxins in mature false smut balls (FSBs) [12]. Numerous studies have revealed that ustiloxins are widely toxic to animals. When fed with contaminated rice grains or feeds for a short time, the domestic animals show a variety of symptoms mainly including diarrhea, hemorrhage, vomiting, ovarian atrophy, and abortion [17,18]. Both ustiloxin A and the crude water extract of FSBs have been reported to cause acute necrosis of liver and kidney in mice [10,14,19,20]. The latest toxicology studies show that ustiloxin A can affect development of early-stage zebrafish [21] and may reduce the population of *Tetrahymena thermophila* by disrupting the cell cycle [22]. Since RFS is becoming increasingly serious in China, ustiloxin pollution has posed a great threat for food safety.

Detection techniques toward mycotoxins are important for the safety of agricultural products. At present, the main detection methods of mycotoxins include instrumental analysis detection and immunoanalysis detection. Instrumental analysis is mainly applied in laboratories, industries, and farmers' enterprises. It is always accurate and reliable, but tends to require expensive instruments and highly-skilled professionals. In recent years, the research on rapid immune detection of small molecule targets has developed rapidly, and immunoassays have been widely used for on-site rapid detection in food security field. The mechanisms of antibody affinity and stability have been newly interpreted [23,24]. Many new antibodies have been developed, such as nanoantibodies and molecular engineering antibodies [25,26]. A number of rapid detection devices and corresponding analysis methods have emerged such as optical and electrochemical immune sensors, new immune kits, and test paper strips [27–29].

To the best of our knowledge, research on the rapid detection technology of ustiloxins is still preliminary. A high-performance liquid chromatographic method for determing ustiloxin A was described, and the limit of detection was 2.5 mg/kg [30]. Then, a HPLC-MS method was developed for simultaneous detection of multiple ustiloxins. The detection limit of ustiloxin A was 5.5 mg/g, while for total ustiloxins it was 0.2 mg/g [31]. In addition to the conventional instrumental detection, the monoclonal antibody enzyme-linked immunoassays against ustiloxin A and B were also reported [32,33]. However, the available monoclonal antibody immunoassays and the corresponding products only target single ustiloxin, thus it is hard to meet the current requirement for simultaneous rapid screening of multiple toxins. Considering this, it is urgent to develop a general rapid detection technology. In the current work, generic hapten against ustiloxins was synthesized and conjugated to protein to produce generic immunogen, which was further injected into mice to generate generic antiserum. The magnetic bead-antibody complex was prepared by immobilizating the general antibody on the surface of magnetic beads. Based on the magnetic bead-antibody complex, an improved ELISA method for the rapid detection of group ustiloxins was established. Our method has great potential to be applied for multi-residue detection of ustiloxins.

2. Results and Discussion

2.1. Design, Synthesis, and Identification of Generic Antigen

Ustiloxins are a group of cyclopep-tides containing a 13-membered cyclic core structure with an ether linkage. The chemical structures of the six known ustiloxins are shown in Figure 1.

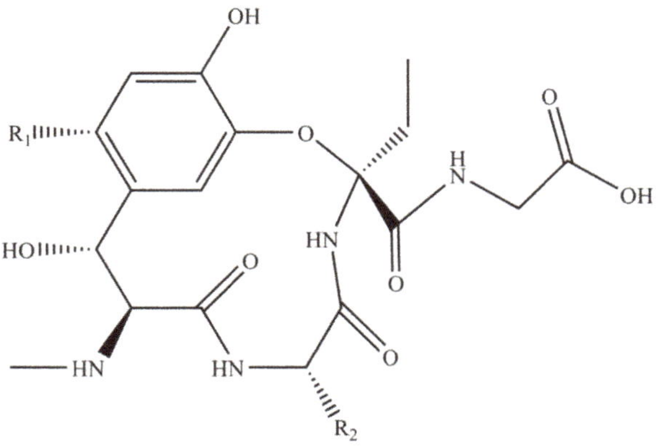

Ustiloxin A
$R_1 = C_5H_7NOS$
$R_2 = CH(CH_2)_2$

Ustiloxin B
$R_1 = C_5H_7NOS$
$R_2 = CH_3$

Ustiloxin C
$R_1 = C_2H_5OS$
$R_2 = CH_3$

Ustiloxin D
$R_1 = H$
$R_2 = CH(CH_2)_2$

Ustiloxin F
$R_1 = H$
$R_2 = CH_3$

Ustiloxin G
$R_1 = C_2H_5OS$
$R_2 = CH(CH_2)_2$

Figure 1. Chemical structures of ustiloxin A, B, C, D, F, and G.

As for the antigen design, we mainly considered the following three design principles in reference to the previously reported method: (1) The group exposed to antigen should be the common moieties, which contributes to high affinity interaction between hapten and antibody; (2) the group exposed to antigen should exhibit similar polarity among analytes; (3) steric hindrance of the substituents should be as small as possible [34].

It should be emphasized that phenyl, amino, and hydroxyl are the molecular basis for the formation of high-affinity interaction between hapten and antibody. According to the above design principle and the structures of the six ustiloxins, the exposure of the common phenyl, amino, and hydroxyl groups of the ustiloxins to the surface of the carrier protein will be conducive to the production of class-specific antibodies. Therefore, we chose the carboxyl to couple with carrier protein so as to promote the exposure of these groups. Additionally, considering the polar similarity of substituents (Residue 1 and Residue 2 in Figure 1) and other factors, we finally chose ustiloxin G as the raw material for antigen synthesis.

The generic antigen was synthesized through active ester method, and the small molecules that did not bind to the carrier proteins were removed. The generic antigen was identified by ultraviolet scanning method. The scanning spectra of the synthesized antigen, the raw material ustiloxin G, and carrier protein BSA (Figure 2) showed that the antigen had two characteristic absorption peaks at 254 nm and 291 nm, that ustiloxin G had a unique absorption peak at 275 nm, and that BSA had a unique absorption peak at 280 nm. These results indicated that the conjugation reaction of the artificial antigen (ustiloxin G-BSA) was successful. Following the same procedures, ustiloxin G-OVA was prepared as a coating antigen.

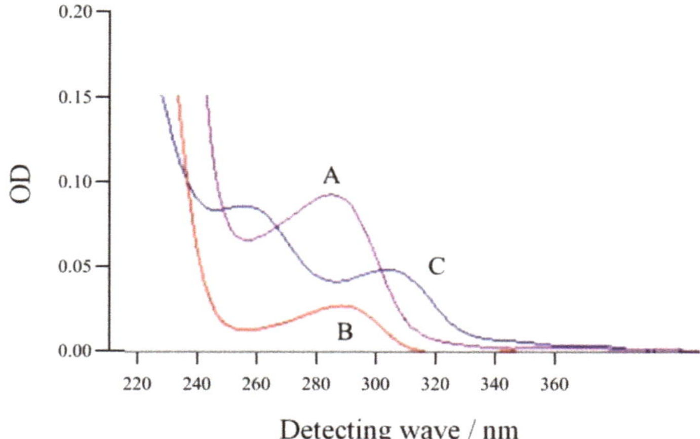

Figure 2. Ultraviolet scanning spectra of ustiloxin G (A), bovine serum albumin (B), and the generic antigen of ustiloxins (C).

2.2. Preparation and Characterization of Generic Antibodies against Ustiloxins

In order to evaluate the above antigen, three mice were immunized with the generic antigen ustiloxin G-BSA by a three-point immune methods. The antisera were collected on the 7th day after each immunization, and the titers were determined by indirect non-competitive ELISA. The titer was expressed as the dilution multiple when the OD value was 1.0. The trends of the titers of the three antisera showed that the polyclonal antibody from mouse 3 had the highest titer (Figure 3). Thus, this polyclonal antibody was selected for the subsequent research.

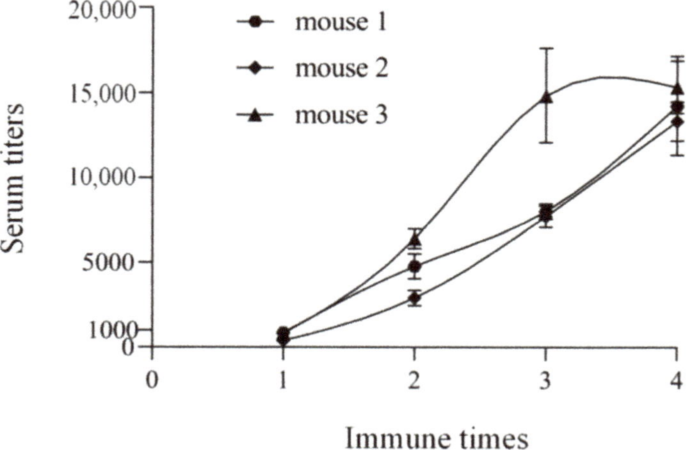

Figure 3. Titer trends of three antisera against generic antigen of ustiloxins.

Using an optimized indirect competitive ELISA, the antibody's sensitivity to ustiloxins was determined, and the result (Table 1) showed that IC_{50} was 0.53 µg/mL for ustiloxin A, 0.34 µg/mL for ustiloxin B, and 0.06 µg/mL for ustiloxin G. The cross-reactivity was 18.4%, 11.9%, and 100% for ustiloxin A, B, and G, respectively.

Table 1. Specificity of generic antibody against ustiloxins.

Analytes	IC$_{50}$ (µg/mL)	Cross-Reactivity (%)
AFB1	no inhibition	0
DON	no inhibition	0
ZEN	no inhibition	0
OTA	no inhibition	0
UST-A	0.53	18.4
UST-B	0.34	11.9
UST-G	0.06	100

2.3. Establishment of Class-Specific Immuno-Magnetic Bead Enzyme-Linked Immunoassay

In order to develop an immuno-magnetic bead enzyme-linked immunoassay, we firstly investigated the time length of magnetic bead activation by glutaraldehyde and the time length of bead-antibody coupling reaction post the bead activation. The shortest coupling time was determined once the OD value reached the maximum. The results indicated that when the magnetic beads were activated by glutaraldehyde for 4 h and then reacted with antibody for 10 min, the OD showed the highest value (Figure 4A). Under this condition, the immuno-magnetic beads of ustiloxins were prepared for further study. According to a classic chessboard titration method [35], the appropriate concentrations of the above immuno-magnetic beads were calculated as 2.0 µg/mL in terms of antibody concentration and that of the coated antigen (ustiloxin G-OVA) was determined as 2.5 µg/mL.

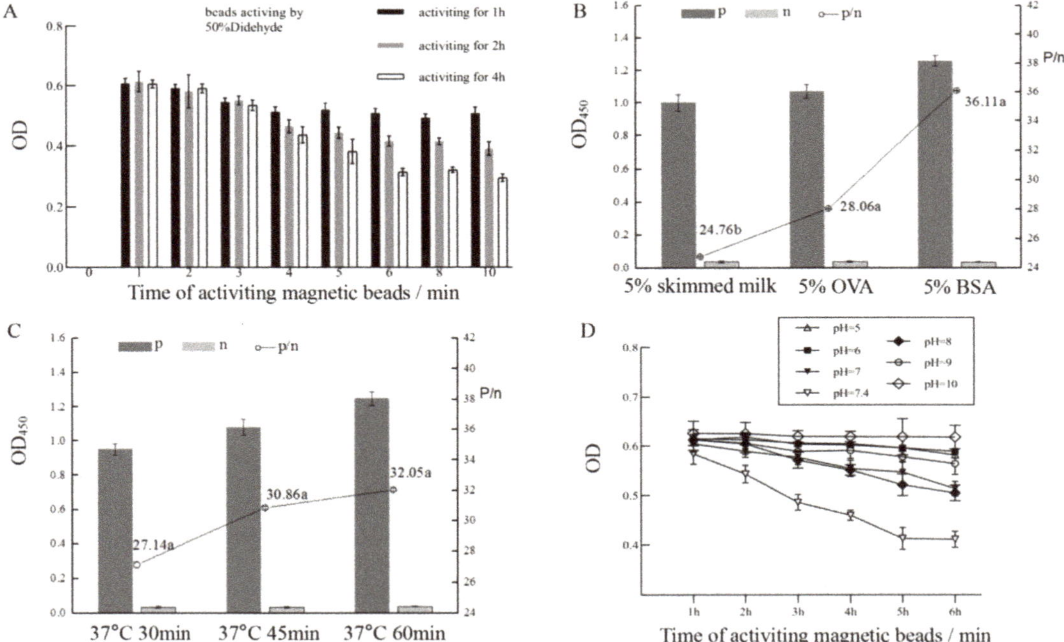

Figure 4. Optimization of immuno-magnetic bead-based enzyme-linked immunoassay conditions including the time length of activation of magnetic beads by glutaraldehyde and the time length of bead-antibody coupling reaction post magnetic activation (**A**), blocking regent types (**B**), antibody reaction time length in ELISA (**C**), and pH values (**D**).

Several factors can affect antibody-antigen binding reaction such as blocking regent types, pH value, and reaction time length. Compared with 5% skimmed milk and 5% OVA, 5% BSA as a blocking reagent was the best selection to prevent nonspecific binding (Figure 4B). Sixty minutes at 37 °C was found to be the optimal reaction time length (Figure 4C). The pH value experiment showed that about 7.4 was the most suitable pH value (Figure 4D).

Finally, based on the above optimized conditions, a generic immuno-magnetic bead enzyme-linked immunoassay method was established for ustiloxin detection. The results indicated that the sensitivity of the established method for ustiloxin A, B, and G was 0.15 µg/mL, 0.14 µg/mL, and 0.04 µg/mL, respectively (Figure 5). It is worth mentioning that our immuno-magnetic bead-based ELISA method obviously improved sensitivity to ustiloxins with the cross-reactivity of ustiloxin A increased to 27.4% and that of ustiloxin B increased to 29.1%.

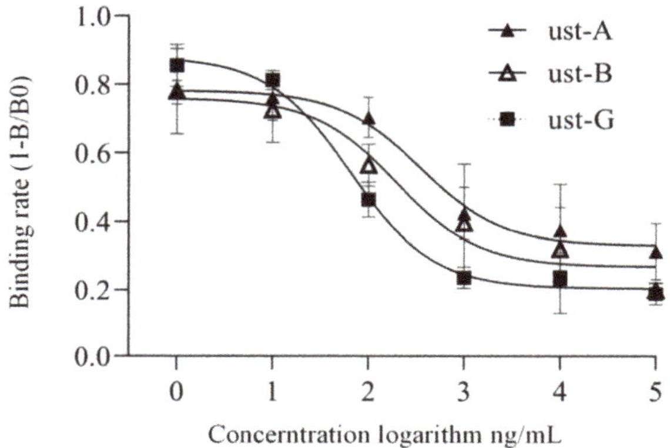

Figure 5. Immuno-magnetic bead-based enzyme-linked immunoassay curves for ustiloxin A, B, and G.

2.4. Evaluation of Accuracy and Standard Deviation

In order to evaluate the accuracy and standard deviation of the developed method above, clean rice was spiked with ustiloxin G as standard. The result (Table 2) showed that the average ustiloxin G recoveries at the high-, mid-, and low-spiked concentration in rice were 101.9%, 107.4%, and 116.4% respectively. These concentration results of our developed method were very close to the theoretical concentrations. As shown in Table 2, the relative standard deviations of our developed ELISA were below 15%, indicating a good repeatability.

We further validated our developed ELISA by using a standard HPLC method and found that the results obtained from both methods were very similar, indicating that the tested results from immuno-magnetic bead-based ELISA could be accepted by the standard HPLC method. Additionally, our result also indicated that the sensitivity of our developed immunoassay was higher than that of standard HPLC.

Table 2. Evaluation of developed method's accuracy and standard deviation using spiked ustiloxin G as standards.

Method	Theoretical Concentration (ng/mL)	Tested Concentration (±SD)/(ng/mL)			Average Recovery (%)
		Rice Brand 1	Rice Brand 2	Rice Brand 3	
Immuno-magnetic bead enzyme immunoassay	0	no detection	no detection	no detection	no calculation
	10	10.96 ± 1.25	12.25 ± 0.92	11.70 ± 1.45	116.4
	50	50.14 ± 4.81	56.09 ± 5.25	54.93 ± 4.38	107.4
	100	101.36 ± 7.56	100.29 ± 8.92	104.00 ± 6.55	101.9
HPLC	10	no detection	no detection	no detection	no calculation
	50	no detection	no detection	no detection	no calculation
	100	105.49 ± 12.43	103.15 ± 7.92	105.81 ± 4.30	104.8

3. Conclusions

In view of the lack of universal rapid screening methods for ustiloxins in rice, we synthesized a generic antigen BSA-ustiloxin G and a coating antigen OVA-ustiloxin. With the conjugate of BSA-ustiloxin G as an immunogen, a generic polyclonal antibody against ustiloxins was obtained and characterized. The result revealed that the generic antigen was suitable for the preparation of generic antibodies against ustiloxins. With the generic antigen and antibody, a sensitive and class-specific immuno-magnetic bead-based enzyme-linked immunoassay was developed for ustiloxin detection, and its sensitivity to ustiloxin A was 0.15 μg/mL, to ustiloxin B was 0.14 μg/mL, and to ustiloxin G was 0.04 μg/mL. The accuracy and repeatability evaluation showed that the recoveries of our developed method ranged from 101.9% to 116.4%, and the new method test result could be accepted by a standard HPLC method. Notably, the sensitivity of the developed method was higher than the standard HPLC, and cross-reactivity of our developed ELISA higher than that of conventional ELISA. Here, we provided a new, sensitive, and generic immunoassay, which can be used for on-site monitoring ustiloxins in rice. The schematic of the assay procedure is shown in Figure 6.

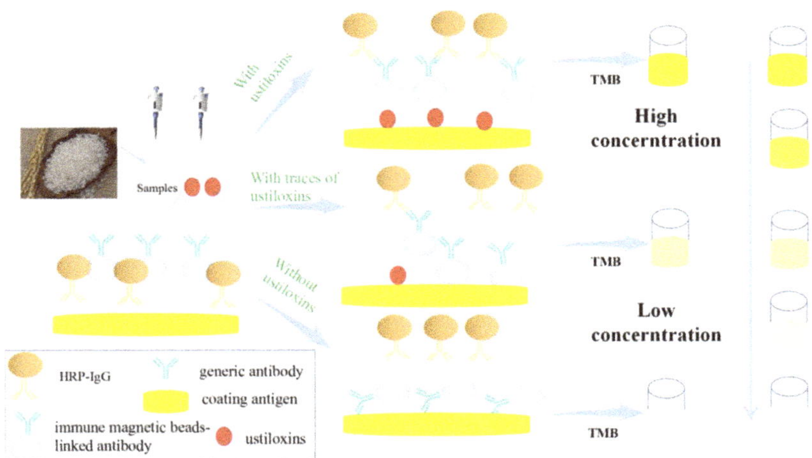

Figure 6. Generic immuno-magnetic bead-based enzyme-linked immunoassay for ustiloxins.

4. Materials and Methods

4.1. Instruments and Experimental Equipment

Costar-clear 96-well culture plates were purchased from Costar (New York, NY, USA). Wellwash 4MK2 automatic plate washer (Thermo, Waltham, MA, USA) and SpectraMax M2e phosphase marker were from Molecular Instruments (Molecules, Los Angeles, CA,

USA). Ultrasonic Cleaner (Jiangsu KH-500E) was from Kunshan Hehuang Ultrasonic Instrument Co., Ltd. (Kunshan, China). Shimadzu Prominence LC-20AT high performance liquid chromatography system (Kyoto, Japan) consisted of two LC-20AT solvent delivery units, one SIL-20A autosampler, one SPD-M20A photodiode array detector, one CBM-20Alite systemcontroller, and one Synergi reversed-phase Hydro-C18 column (250 mm, 4.6 mm, 10 mm) (Phenomenex, Torrance, CA, USA).

4.2. Materials and Reagents

The mycotoxins ustiloxin A (UST-A), ustiloxin B (UST-B), and ustiloxin G (UST-G) used in this study were prepared in our laboratory, according to the previously reported method [34,35]. 1-ethyl-(3-dimethylaminopropyl)-carbodiimide hydrochloride (EDC), N-hydroxysuccinimide (NHS), Freund's incomplete adjuvant (FIA, Freund's incomplete adjuvant), Freund's complete adjuvant (FCA), bovine serum albumin (BSA), ovalbumin (OVA), 3, 3′, 5, 5′-tetramethylbenzidine (TMB), standards of aflatoxin B1 (AFB1), deoxynivalenol (DON), zearalenone (ZEA), and ochratoxin (OTA) were purchased from Sigma-Aldrich (St. Louis, MO, USA). Horseradish peroxidase-labeled goat anti-mouse IgG antibody (HRP-IgG) was purchased from Wuhan Boster Biological Co., Ltd. (Wuhan, China). Magnetic beads (700 nm, with $-NH_2$) were purchased from Hangzhou Kunteng Nano Technology Co., Ltd. (Hangzhou, China).

Phosphate buffer (PBS) was prepared as follows. NaCl (16 g), KCl (0.4 g), Na_2HPO_4 $12H_2O$ (5.8 g), and KH_2PO_4 (0.4 g) were added simultaneously into a volumetric flask, and then deionized water was added to reach a final volume of 2000 mL. The coating solution was formulated as follows. Na_2CO_3 (3.18 g) and $NaHCO_3$ (5.86 g) were added together into a volumetric flask, and then deionized water was added to obtain carbonate buffer with a volume of 2000 mL and concentration of 0.05 mol/L. ELISA detergent was prepared as follows. One milliliter tween-20 was dissolved in 2 L PBS buffer (0.01 mol/L) and mixed well to obtain 0.05% PBST. Citric acid buffer was obtained by the following procedures. $Na_2HPO_4 \cdot 12H_2O$ (9.205 g) and citric acid (4.665 g) were added into deionized water to obtain 500 mL solution, which was stored at 4 °C. Urea peroxide (3.0 g) was dissolved into 100 mL anhydrous ethanol to obtain urea peroxide solution, which was stored at 4 °C. The 0.2 g TMB was dissolved into 100 mL anhydrous ethanol to obtain TMB solution, and it was stored at −20 °C. The 0.5 mL TMB solution, 32 μL urea peroxide solution, and 9.5 mL citric acid buffer were mixed to obtain TMB color solution. The 11 mL concentrated sulfuric acid was dissolved into 89 mL deionized water to obtain stopping coloring solution, and it was stored at room temperature.

4.3. Preparation of Immunogen and Coating Antigen

The 1.0 mg Ustiloxin G was dissolved in 1.0 mL N,N-dimethylformamide (DMF), and then 2.485 mg N-hydroxysuccinimide (NHS) and 1.0 mg 1-ethyl-(3-dimethylaminopropyl)-carbodiimidehydrochloride (EDC) were added and mixed well. The mixture was placed at room temperature and stirred at 150 r/min for 2 h. The supernatant containing the active ester of ustiloxin G was finally obtained for subsequent experiments.

The 1.0 mg carrier protein (BSA or OVA) was dissolved in 1.0 mL PBS (0.01 mol/L), and then the active ester solution of ustiloxin G was added drop by drop into the protein solution when stirring. The mixture was then placed into a shaker at 4 °C and shaken at 200 r/min overnight. Finally, those small molecules unbound to protein were removed by ultrafiltration and centrifugation at 4000 g/min for 30 min, and the conjugates were retained in the ultrafiltration tube and were re-dissolved with PBS.

4.4. Preparation of Polyclonal Antibody against Ustiloxins

In the initial immunization, 2 mg USTG-BSA conjugate was dissolved in a sterilized 0.5 mL NaCl solution (0.9%) and then emulsified with an equal volume of FCA. Three 6-week-old female BALB/c mice were immunized by multiple-point subcutaneous injection with the final water-in-oil emulsion described above. Booster injections were performed with an equal volume of the FIA-emulsified antigen three times at 3-week intervals. To investigate the immune response to immunogen, the antisera were collected from the tails of the four mice at day 9 post immunization and assayed with ustiloxin G-OVA by indirect noncompetitive ELISA.

To examine the sensitivity and specificity of antibodies, the inhibition experiments of mycotoxin AFB1; DON; ZEA; OTA; and UST-A, B, and G were conducted using a traditional indirect competitive ELISA, and their 50% inhibition concentrations (IC_{50}) were calculated by the classic four-parameter equation [27]. The specificity was usually expressed as cross-reactivity, which was calculated according to the following formula.

$$\text{Cross-reactivity} = (IC_{50} \text{ UST-G}/IC_{50} \text{ UST-A, B, or other mycotoxin}) * 100\%$$

Briefly, the procedures of the traditional indirect competitive ELISA were as follows: (1) The plates were coated with 100 mL/well USTG-OVA at an appropriate concentration in 0.05 M PBS (pH 7.4) and stood for 2 h at 37 °C (2). After washing three times with 300 mL 0.05% PBST, 200 mL of 5% OVA in the PBST solution was added to each well and incubated for 1 h at 37 °C. (3) After another three washes, the 100 mL/well generic polyclonal antibody against ustiloxins at an appropriate solution was added into each well of the plates; (4) after 1-h incubation at 37 °C, the plates were rewashed, IgG-HRP (diluted at 1/8000 in the PBST, 100 mL/well) was added, and then the plates were incubated for 45 min at 37 °C; (5). After six washes, the color was developed by adding 100 mL freshly prepared substrate solution (containing 9.5 mL phosphate-citrate buffer (pH 5.0) and 0.5 mL TMB (2 mg/mL) dissolved in ethanol), and 32 mL urea-hydrogen peroxide (3%, *w/v*)), and the mixture was incubated for 15 min at 37 °C in the dark; (6) the 50 mL of the stopping coloring solution (H_2SO_4) was added to each well, and the absorbance at 450 nm was measured with a microplate reader.

4.5. Preparation of Immuno-Magnetic Beads

The solvent of the magnetic bead solution (5 mg in 2 mL) was removed by magnetic separation, and magnetic beads were washed with 5 mL PBS solution (0.05 M) three times. Then 100 mL glutaraldehyde solution (50%) was added to the washed beads. The mixture was shaken for 4 h at room temperature.

After the activation, the magnetic beads were magnetically separated and washed three times with PBS solution. The beads were then added into the generic antibody solution (0.5 mg in 2.5 mL PBS) for conjugation for 10 min. The sites where the beads were unbound with the antibody were then blocked with 5% BSA for 1 h at room temperature. Finally, the conjugates were magnetically separated for subsequent use, and the residue antibodies in the reaction solution were tested by indirect noncompetitive ELISA.

4.6. Optimization of Generic Immuno-Magnetic Bead-Based Enzyme-Linked Immunoassay

OD value of indirect noncompetitive ELSA was used to evaluate immunoassay performance and to reveal the influence of multiple factors on reaction of the antigen–antibody complex. Multiple factors were investigated such as time length of the activation (1, 2, and 4 h) and the conjugation (1, 2, 3, 4, 5, 6, 8, and 10 min) between the immunomagnetic beads and antibody, blocking regent (5% OVA, 5% BSA, and 5% skim milk powder), time length of competitive reaction (30, 45, and 60 min) in the immuno-magnetic bead enzyme-linked immunoassay procedures, and pH (5.0, 6.0, 7.0, 7.4, 8.0, 9.0, and 10.0) of the competitive reaction solution.

It should be noted that the only difference between immuno-magnetic bead-based ELISA and the traditional ELISA lay in that the generic antibody was used directly for the traditional ELISA, whereas the immunomagnetic beads were used for the immuno-magnetic bead-based ELISA.

4.7. Evaluation on Accuracy and Repeatability of Usitloxin G Recovery

Three brands of clean rice uncontaminated by ustiloxin were collected from the local market. Ustiloxin G was used as the standard to evaluate the accuracy of the developed immunoassay. In the experiments, the uncontaminated rice was spiked with ustiloxin G, and then the spiked samples were ground. The ground samples were sonicated with ultrapure water for 4 h at room temperature. After centrifugation, the supernatant of each sample was filtered with a 0.22 μm microporous filter and transferred into a new tube before analysis. Theoretically, ustiloxin G at three final concentrations (10, 50, and 100 ng/mL) should be obtained. The extracts were finally detected by our developed immuno-magnetic bead enzyme immunoassay method and conventional HPLC method described previously [36,37].

Author Contributions: Conceptualization, Y.H., H.L. and Q.Z.; methodology, X.T. and Q.Z.; validation, J.H., H.L. and Q.Z.; formal analysis, L.Z. and J.H.; resources, H.L. and Q.Z.; data curation, L.Z. and Q.Z.; writing—original draft preparation, Y.H.; writing—review and editing, X.T., H.L. and Q.Z.; project administration, Q.Z. and H.L.; funding acquisition, Q.Z. and H.L. All authors have read and agreed to the published version of the manuscript.

Funding: This research was funded by the National Natural Science Foundation of China (31801678) and the Key Research & Development Program of Hubei Province (2021BBA236).

Institutional Review Board Statement: Not applicable.

Informed Consent Statement: Not applicable.

Conflicts of Interest: The authors declare no conflict of interest.

References

1. Sun, W.; Fan, J.; Fang, A.; Li, Y.; Tariqjaveed, M.; Li, D.; Hu, D.; Wang, W.M. *Ustilaginoidea virens*: Insights into an emerging rice pathogen. *Annu. Rev. Patytopathol.* **2020**, *58*, 363–385. [CrossRef]
2. Chen, X.; Li, X.; Li, P.; Chen, X.; Liu, H.; Huang, J.; Luo, C.; Hsiang, T.; Zheng, L. Comprehensive identification of lysine 2-hydroxyisobutyrylated proteins in *Ustilaginoidea virens* reveals the involvement of lysine 2-hydroxyisobutyrylation in fungal virulence. *J. Integr. Plant Biol.* **2021**, *63*, 409–425. [CrossRef]
3. Tang, J.; Chen, X.; Yan, Y.; Huang, J.; Luo, C.; Hsiang, T.; Zheng, L. Comprehensive transcriptome profiling reveals abundant long non-coding RNAs associated with development of the rice false smut fungus, *Ustilaginoidea virens*. *Environ. Microbiol.* **2021**, *23*, 4998–5013. [CrossRef] [PubMed]
4. Zhang, Y.; Zhang, K.; Fang, A.; Han, Y.; Yang, J.; Xue, M.; Bao, J.; Hu, W.; Zhou, B.; Sun, X.; et al. Specific adaptation of *Ustilaginoidea virens* in occupying host florets revealed by comparative and functional genomics. *Nat. Commun.* **2014**, *5*, 3849. [CrossRef] [PubMed]
5. Song, J.; Wei, W.; Lv, B.; Lin, Y.; Yin, W.; Peng, Y.; Schnabel, G.; Huang, J.; Jiang, D.; Luo, C. Rice false smut fungus hijacks the rice nutrients supply by blocking and mimicking the fertilization of rice ovary. *Environ. Microbiol.* **2016**, *18*, 3840–3849. [CrossRef] [PubMed]
6. Zhou, Y.; Xie, X.; Zhang, F.; Wang, S.; Liu, X.; Zhu, L.; Xu, J.; Gao, Y.; Li, Z. Detection of quantitative resistance loci associated with resistance to rice false smut (*Ustilaginoidea virens*) using introgression lines. *Plant Pathol.* **2013**, *63*, 365–372. [CrossRef]
7. Lu, D.; Yang, X.; Mao, J.; Ye, H.; Wang, P.; Chen, Y.; He, Z.; Chen, F. Characterising the pathogenicity diversity of *Ustilaginoidea virens* in hybrid rice in China. *J. Plant Pathol.* **2009**, *91*, 443–451.
8. Yu, M.; Yu, J.; Hu, J.; Huang, L.; Wang, Y.; Yin, X.; Nie, Y.; Meng, X.; Wang, Y.; Liu, Y. Identification of pathogenicity-related genes in the rice pathogen *Ustilaginoidea virens* through random insertional mutagenesis. *Fungal Genet. Biol.* **2015**, *76*, 10–19. [CrossRef]
9. Tang, J.; Zheng, L.; Jia, Q.; Liu, H.; Hsiang, T.; Huang, J. PCR markers derived from comparative genomics for detection and identification of the rice pathogen *Ustilaginoidea virens* in plant tissues. *Plant Dis.* **2017**, *101*, 1515–1521. [CrossRef] [PubMed]
10. Nakamura, K.; Izumiyama, N.; Ohtsubo, K.; Koiso, Y.; Iwasaki, S.; Sonoda, R.; Fujita, Y.; Yaegashi, H.; Sato, Z. Lupinosis in mice caused by ustiloxin and a crude extract of fungal culture of *Ustilaginoidea virens*. *JSM Mycotoxins* **1992**, *35*, 41–43. [CrossRef]
11. Koiso, Y.; Li, Y.; Iwasaki, S. Ustiloxins, antimitotic cyclic peptides from false smut balls on rice panicles caused by *Ustilaginoidea virens*. *J. Antibiot.* **1994**, *47*, 765–773. [CrossRef]
12. Wang, X.; Fu, X.; Lin, F.; Sun, W.; Meng, J.; Wang, A.; Lai, D.; Zhou, L.; Liu, Y. The contents of ustiloxins A and B along with their distribution in rice false smut balls. *Toxins* **2016**, *8*, 262. [CrossRef] [PubMed]
13. Hu, Z.; Zheng, L.; Huang, J.; Zhou, L.; Liu, C.; Liu, H. Ustiloxins A is produced early in experimental *Ustilaginoidea virens* infection and affects transcription in rice. *Curr. Microbiol.* **2020**, *77*, 2766–2774. [CrossRef] [PubMed]
14. Koiso, Y.; Natori, M.; Iwasaki, S. Ustiloxin: A phytotoxin and a mycotoxin from false smut balls on rice panicles. *Tetrahedron Lett.* **1992**, *33*, 4157–4160. [CrossRef]

15. Koiso, Y.; Morisaki, N.; Yamashita, Y. Isolation and structure of an antimitotic cyclic peptide, ustiloxin F: Chemical interrelation with a homologous peptide, ustiloxin B. *J. Antibiot.* **1998**, *51*, 418–422. [CrossRef] [PubMed]
16. Wang, X.; Wang, J.; Lai, D.; Wang, X.; Dai, J.; Zhou, L.; Liu, Y. Ustiloxin G, a new cyclopeptide mycotoxin from rice false smut balls. *Toxins* **2017**, *9*, 54. [CrossRef] [PubMed]
17. Li, B. Prevention and control of pig feed poisoning caused by ustiloxins. *Fujian J. Anim. Hus. Vet. Med.* **2005**, *27*, 70–71.
18. Wu, Y. Investigation and prevention report of ustiloxins poisoning in livestock. *Chin. J. Tradit. Vet. Sci.* **2004**, *5*, 20–22.
19. Nakamura, K.; Izumiyama, N.; Ohtsubo, K. Apoptosis induced in the liver, kidney and urinary bladder of mice by the fungal toxin produced by *Ustilaginoidea virens*. *Mycotoxins* **1993**, *38*, 25–30. [CrossRef]
20. Nakamura, K.; Izumiyama, N.; Ohtsubo, K.; Koiso, Y.; Iwasaki, S.; Sonoda, R.; Fujita, Y.; Yaegashi, H.; Sato, Z. "Lupinosis"-Like lesions in mice caused by ustiloxin, produced by *Ustilaginoieda virens*: A morphological study. *Nat. Toxins* **1994**, *2*, 22–28. [CrossRef] [PubMed]
21. Hu, Z.; Dang, Y.; Liu, C.; Zhou, L.; Liu, H. Acute exposure to ustiloxin A affects growth and development of early life zebrafish, *Danio rerio*. *Chemosphere* **2019**, *226*, 851–857. [CrossRef] [PubMed]
22. Cheng, S.; Liu, H.; Sun, Q.; Kong, R.; Letcher, R.; Liu, C. Occurrence of ustiloxin A in surface waters of paddy fields in Enshi, Hubei, China, and its toxicity in *Tetrahymena Thermophila*. *Environ. Pollout.* **2019**, *251*, 901–909. [CrossRef]
23. Sun, S.; Sen, S.; Kim, N.; Magliery, T.; Schultz, P.; Wang, F. Mutational analysis of 48G7 reveals that somatic hypermutation affects both antibody stability and binding affinity. *J. Am. Chem. Soc.* **2013**, *135*, 9980–9983. [CrossRef] [PubMed]
24. Wang, F.; Sen, S.; Zhang, Y.; Ahmad, I.; Zhu, X.; Wilson, I.; Smider, V.; Magliery, T.; Schultz, P. Somatic hypermutation maintains antibody thermodynamic stability during affinity maturation. *Proc. Natl. Acad. Sci. USA* **2013**, *110*, 4261–4266. [CrossRef] [PubMed]
25. Cheung, W.; Beausoleil, S.; Zhang, X.; Sato, S.; Schieferl, S.; Wieler, J.; Beaudet, J.; Ramenani, R.; Popova, L.; Comb, M.; et al. A proteomics approach for the identification and cloning of monoclonal antibodies from serum. *Nat. Biotechnol.* **2012**, *30*, 447–452. [CrossRef]
26. Da Rosa, S.T.; Rossotti, M.; Carleiza, C.; Carrión, F.; Pritsch, O.; Ahn, K.; Last, J.; Hammock, B.; Gonzalez-Sapienza, G. Competitive selection from single domain antibody libraries allows isolation of high-affinity antihapten antibodies that are not favored in the llama immune response. *Anal. Chem.* **2011**, *83*, 7213–7220. [CrossRef]
27. Dzantiev, B.; Byzova, N.; Urusov, A.; Zherdev, A. Immunochromatographic methods in food analysis. *TrAC Trend. Anal. Chem.* **2014**, *55*, 81–93. [CrossRef]
28. Zhang, W.; Asiri, A.; Liu, D.; Du, D.; Lin, Y. Nanomaterial-based biosensors for environmental and biological monitoring of organophosphorus pesticides and nerve agents. *TrAC Trend. Anal. Chem.* **2014**, *54*, 1–10. [CrossRef]
29. He, Q.; Xu, Y.; Zhang, C.; Li, Y.; Huang, Z. Phage-borne peptidomimetics as immunochemical reagent in dot-immunoassay for mycotoxin zearalenone. *Food Control* **2014**, *39*, 56–61. [CrossRef]
30. Miyazaki, S.; Matsumoto, Y.; Uchihara, T.; Morimoto, K. High-Performance liquid chromatographic determination of ustiloxin A in forage rice silage. *J. Vet. Med. Sci.* **2009**, *71*, 239–241. [CrossRef] [PubMed]
31. Cao, Z.; Sun, L.; Mou, R.; Lin, X.; Zhou, R.; Ma, Y.; Chen, M. Analysis of ustiloxins in rice using polymer cation exchange cleanup followed by liquid chromatography-tandem mass spectrometry. *J. Chromatogr. A* **2016**, *1476*, 46–52. [CrossRef]
32. Fu, X.; Wang, W.; Li, Y.; Wang, X.; Tan, G.; Lai, D.; Wang, M.; Zhou, L.; Wang, B. Development of a monoclonal antibody with equal reactivity to ustiloxins A and B for quantification of main cyclopeptide mycotoxins in rice samples. *Food Control* **2018**, *92*, 201–207. [CrossRef]
33. Fu, X.; Wang, X.; Cui, Y.; Wang, A.; Lai, D.; Liu, Y.; Li, Q.; Wang, B.; Zhou, L. A monoclonal antibody-based enzyme-linked immunosorbent assay for detection of ustiloxin A in rice false smut balls and rice samples. *Food Chem.* **2015**, *81*, 140–145. [CrossRef] [PubMed]
34. Zhang, Q.; Wang, L.; Ahn, K.; Sun, Q.; Hu, B.; Wang, J.; Liu, F. Hapten heterology for a specific and sensitive immunoassay for fenthion. *Anal. Chim. Acta* **2007**, *596*, 303–311. [CrossRef] [PubMed]
35. Li, P.; Zhang, Q.; Zhang, W. Immunoassays for aflotoxins. *Trends Anal. Chem.* **2009**, *28*, 1115–1126. [CrossRef]
36. Shan, T.; Sun, W.; Liu, H.; Gao, S.; Lu, S.; Wang, M.; Sun, W.; Chen, Z.; Wang, S.; Zhou, L. Determination and analysis of ustiloxins A and B by LC-ESI-MS and HPLC in false smut balls of rice. *Int. J. Mol. Sci.* **2012**, *13*, 11275–11287. [CrossRef]
37. Shan, T.; Sun, W.; Wang, X.; Fu, X.; Sun, W.; Zhou, L. Purification of ustiloxins A and B from rice false smut balls by macroporous resins. *Molecules* **2013**, *18*, 8181–8199. [CrossRef] [PubMed]

Article

The Regulatory Mechanism of Water Activities on Aflatoxins Biosynthesis and Conidia Development, and Transcription Factor AtfB Is Involved in This Regulation

Longxue Ma [1,†], Xu Li [1,†], Xiaoyun Ma [1], Qiang Yu [2], Xiaohua Yu [2], Yang Liu [3], Chengrong Nie [3], Yinglong Zhang [4,*] and Fuguo Xing [1,*]

1   Institute of Food Science and Technology, Chinese Academy of Agricultural Sciences, Beijing 100193, China; longxuem@foxmail.com (L.M.); lixu@caas.cn (X.L.); xiaoyunma29@foxmail.com (X.M.)
2   Qingdao Tianxiang Foods Group Co., Qingdao 266737, China; yuqiang@tianxianggroup.cn (Q.Y.); yuxiaohua@tianxianggroup.cn (X.Y.)
3   School of Food Science and Engineering, Foshan University, Foshan 528231, China; liuyang@fosu.edu.cn (Y.L.); niecr@126.com (C.N.)
4   Shandong Institute of Commerce and Technology, Jinan 250103, China
*   Correspondence: zhangyinglong@sict.edu.cn (Y.Z.); xingfuguo@caas.cn (F.X.)
†   These authors contributed equally to this work.

Citation: Ma, L.; Li, X.; Ma, X.; Yu, Q.; Yu, X.; Liu, Y.; Nie, C.; Zhang, Y.; Xing, F. The Regulatory Mechanism of Water Activities on Aflatoxins Biosynthesis and Conidia Development, and Transcription Factor AtfB Is Involved in This Regulation. *Toxins* **2021**, *13*, 431. https://doi.org/10.3390/toxins13060431

Received: 23 March 2021
Accepted: 18 June 2021
Published: 21 June 2021

Abstract: Peanuts are frequently infected by *Aspergillus* strains and then contaminated by aflatoxins (AF), which brings out economic losses and health risks. AF production is affected by diverse environmental factors, especially water activity (a_w). In this study, *A. flavus* was inoculated into peanuts with different a_w (0.90, 0.95, and 0.99). Both AFB_1 yield and conidia production showed the highest level in a_w 0.90 treatment. Transcriptional level analyses indicated that AF biosynthesis genes, especially the middle- and later-stage genes, were significantly up-regulated in a_w 0.90 than a_w 0.95 and 0.99. AtfB could be the pivotal regulator response to a_w variations, and could further regulate downstream genes, especially AF biosynthesis genes. The expressions of conidia genes and relevant regulators were also more up-regulated at a_w 0.90 than a_w 0.95 and 0.99, suggesting that the relative lower a_w could increase *A. flavus* conidia development. Furthermore, transcription factors involved in sexual development and nitrogen metabolism were also modulated by different a_w. This research partly clarified the regulatory mechanism of a_w on AF biosynthesis and *A. flavus* development and it would supply some advice for AF prevention in food storage.

Keywords: water activity; aflatoxin biosynthesis; conidia development; regulatory mechanism; AtfB

Key Contribution: This research revealed the regulatory mechanism of a_w on AF biosynthesis and *A. flavus* development, and transcription factor AtfB is involved in the regulation. These results will provide some possible targets for AF prevention in food storage.

1. Introduction

Peanut is an important economical crop for oil production and nutritious addition in human consumption. However, aflatoxigenic *Aspergillus* strains infection and aflatoxins (AF) contamination bring out immense human health risks and huge economic losses for the peanut industry. AF are the polyketide-derived furanocoumarins with strong carcinogenicity that associated with both acute and chronic toxicity for animals and humans [1]. More than 28% hepatocellular carcinoma cases are induced by AF contamination in the world [2]. Among the diverse AF, aflatoxin B_1 (AFB_1), as the most toxic and dangerous one, is usually high-level-produced by some aflatoxigenic *Aspergillus* strains [3]. Therefore, investigating *A. flavus* growth and metabolism, especially AF biosynthesis, is extremely essential for controlling AF contamination.

The AF biosynthesis and fungal development of *A. flavus* are affected by diverse environmental factors, such as water activity (a_w), temperature, pH, carbon source, nitrogen source, and oxidative stress. Based on the definition of U. S. Food and Drug Administration (FDA), a_w of a food is the ratio between the vapor pressure of the food itself, when in a completely undisturbed balance with the surrounding air media, and the vapor pressure of distilled water under identical conditions. So, a_w as a parameter to measure the freely available water in food or substrate is directly related to the food microbial growth in a specific condition [4]. More importantly, a_w was regarded as a central environmental factor, and could co-modulate the fungal development and toxin production of *Aspergillus* spp. with other environmental factors [5–7]. Previous studies reported that the proper a_w conditions for AF biosynthesis were dependent on the other environmental factors, for example, temperature, pH, light, and especially culture substrates [5,8,9]. However, few researchers focused on the effect of peanut substrates with different a_w on *A. flavus* development and AF production.

As the most important characters of *A. flavus*, AF biosynthesis has been well researched in past decades. More than 20 structural genes, located in the 80-kb AF cluster, are involved in the series enzymatic reactions, and transform acetyl-CoA to AFB_1, AFB_2, AFG_1, and AFG_2 [10]. Two pathway specific regulators, DNA binding protein AflR and transcriptional co-activator AflS, are affected by other regulators or environmental factors, and then modulate the structural genes' transcriptions [9,11]. AF production are also regulated by plenty of global regulators including the velvet complex, MAPK pathway factors, oxidative-stress-related regulators, G-protein receptors, oxylipin proteins, as well as many oxidative stress transcription factors (TFs) [10,12]. All AF biosynthetic enzymes and AF regulators constitute an extremely complicated system, and diverse environmental factors affect AF production by adjusting the expression of the AF regulatory system. In previous studies, the expression of AF structural genes could have been affected by diverse a_w, and the ratio of *aflS*/*aflR* was more down-regulated in a_w 0.99 than a_w 0.96 [6,8,9,13]. However, the mechanism of a_w on AF biosynthesis regulation is still unclear.

Transcriptome analysis is regarded as an effective and efficient method to discover the new regulatory mechanisms. In previous studies, the optimal a_w for AF biosynthesis were in the range of 0.90–0.99 at the different environmental combinations [6,8,9,13]. In this study, the a_w of shelled peanuts were adjusted as 0.90, 0.95, 0.99, and the AF production and fungal growth were confirmed at different a_w. By comprehensive transcriptional analysis, AF cluster genes, conidia development genes, and several TFs were significantly up-regulated at a_w 0.90, and AtfB was regarded as the critical TFs for AF regulation in diverse a_w. This work contributes to better understanding of the regulatory mechanism of a_w on *A. flavus* development and AF biosynthesis, and it is helpful to reduce the AF contamination in peanuts storage.

2. Results

2.1. Water Activity Affects the Conidia Production and the AFB_1 Production of A. flavus in Peanuts

After 10 days cultivation, almost all of the peanuts at a_w 0.90 were covered by the green conidia and mycelia, while the conidia and the green color were significantly less at a_w 0.95 (Figure 1A,B). At a_w 0.99, peanuts were only coated by white mycelia, but without obvious conidia production (Figure 1A,B). After counting the peanut-washed suspensions by hemocytometer, the conidia concentrations were more than 3800 conidia/mL in a_w 0.90, and less than 800 conidia/mL in a_w 0.95, but few conidia were in a_w 0.99 treatment (Figure 1C). The AFB_1 levels in contaminated peanuts in different a_w treatments were also examined (Figure 1D). At a_w 0.90, 568 µg/g AFB_1 were detected, while AFB_1 levels were significantly decreased at a_w 0.95 and 0.99, with 212 µg/g and 36 µg/g, respectively (Figure 1D). So, these results concluded that in shelled peanuts with a_w 0.90–0.99, the conidia development and AFB_1 production of *A. flavus* were increased in the relatively lower a_w conditions.

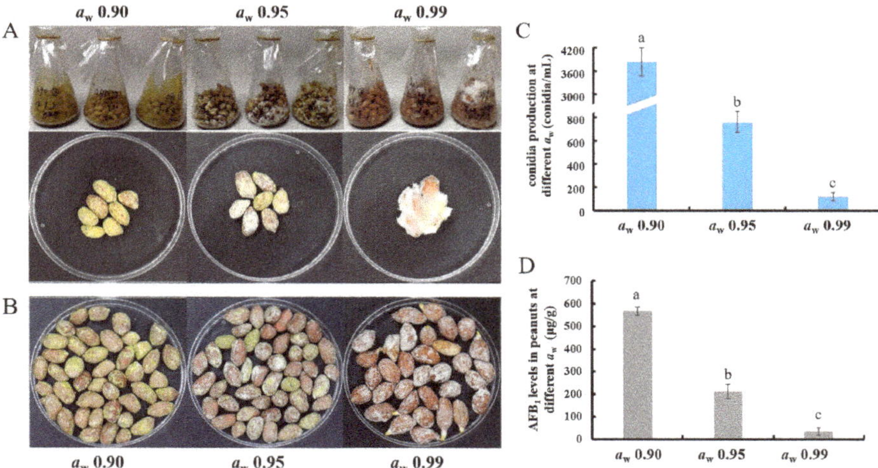

Figure 1. The differences of AFB$_1$ yield and conidia production in peanuts in different a_w. (**A**) The inoculated peanuts with different a_w were placed in flasks for 10 days' cultivation, and (**B**) 25 g treated peanuts were used for AFB$_1$ detection and conidia examination. (**C**) *A. flavus* conidia from peanuts were counted by hemocytometer, and (**D**) AFB$_1$ levels in different a_w peanuts were detected by HPLC. All experiments were performed in three independent biological replicates, and results were represented as means ± SD. Samples marked with different letters show a significant difference at $p < 0.05$.

2.2. Transcriptome Analyses of the A. flavus Genes Expressions in Different Water Activity

To explore the regulatory mechanisms of a_w on *A. flavus* development and AF biosynthesis in peanuts, transcriptome analyses were performed. A total of 14,472 genes were mapped to the *A. flavus* NRRL3357 genome and 671 novel genes were identified from the transcriptome data. Compared with a_w 0.95 treatment, 834 DEGs of *A. flavus* in a_w 0.90 were up-regulated, while 148 DEGs were down-regulated (Figure 2A). A total of 2667 DEGs with 1760 up-regulated and 907 down-regulated were identified in a comparison of a_w 0.90 vs. 0.99 (Figure 2B). In a comparison of a_w 0.95 vs. 0.99, 233 genes were increased, and 95 genes were decreased (Figure 2C). A heat map of the DEGs clustering also showed the obviously differential expression pattern among the three a_w conditions, of which the most genes were up-regulated in a_w 0.90 treatment, while two thirds of the genes were down-regulated at a_w 0.99 (Figure 2D). GO annotation analysis of the comparisons of a_w 0.90 vs. 0.95 and a_w 0.90 vs. 0.99 found that DEGs were enriched in oxidation-reduction process and transmembrane transport in biological process, the intrinsic component of the membrane, the integral component of the membrane, the membrane part, the membrane in the cellular component, and catalytic activity in molecular function (Figure 3A,B). DEGs in a_w 0.95 vs. 0.99 were enriched in similar GO items, such as oxidation-reduction process, single-organism transport, transmembrane transport in biological process, the intrinsic component of membrane, the integral component of the membrane in the cellular component, and oxidoreductase activity in molecular function (Figure 3C). KEGG pathway annotation revealed DEGs of the different a_w comparisons were mainly enriched in biosynthesis of secondary metabolites, steroid biosynthesis, nitrogen metabolism, ribosome, valine, leucine and isoleucine degradation, and starch and sucrose metabolism (Figure 3D–F).

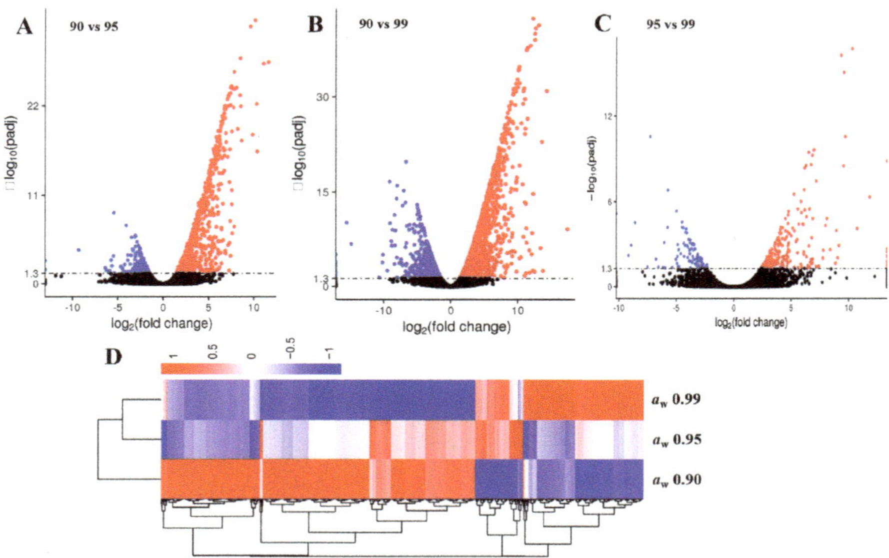

Figure 2. Transcriptomic analyses of *Aspergillus flavus* in different a_w. The volcano plots of the pairwise comparisons in (**A**) a_w 0.90 vs. 0.95, (**B**) a_w 0.90 vs. 0.99, and (**C**) a_w 0.95 vs. 0.99. Up-regulated and down-regulated genes were showed with red spots and blue spots, respectively, and no significantly changed genes were presented with black spots. (**D**) Cluster analysis of DEGs in diverse a_w. Up-regulated and down-regulated genes were represented in red and blue, respectively. The transcriptomic analyses were performed in three independent biological replicates.

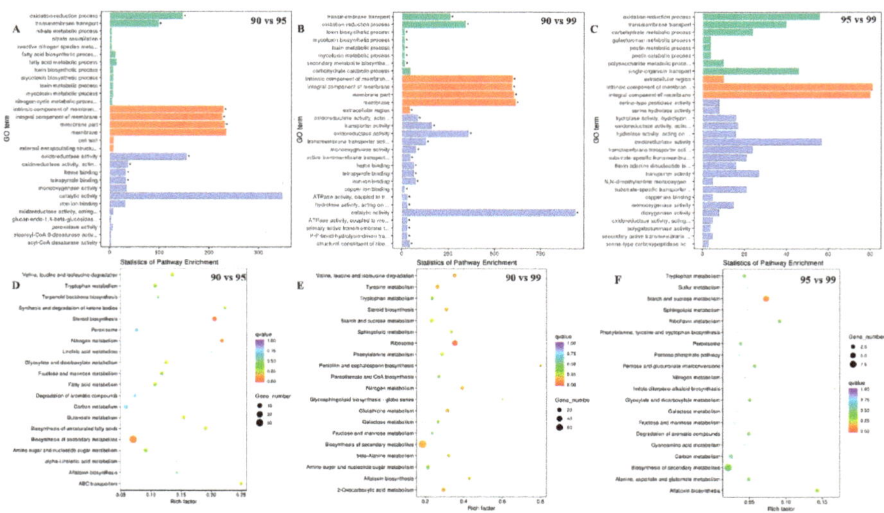

Figure 3. GO annotation and KEGG enrichment of DEGs in different a_w. Bar charts demonstrated the GO-enriched results in comparisons of (**A**) a_w 0.90 vs. 0.95, (**B**) a_w 0.90 vs. 0.99, and (**C**) a_w 0.95 vs. 0.99. The number of enriched genes and the names of GO terms are showed in *X*-axis and *Y*-axis, respectively. Biological process, cellular components, and molecular function were represented by the green bars, orange bars, and blue bars, respectively. The top 20 enriched KEGG pathways were showed in (**D**) a_w 0.90 vs. 0.95, (**E**) a_w 0.90 vs. 0.99, and (**F**) a_w 0.95 vs. 0.99. The rich factors and the pathway names are showed in *X*-axis and *Y*-axis, respectively. The size of spots represented the number of enriched genes, and different colors described the *q*-value.

2.3. Expression Changes of AF Cluster Genes in Different a_w Conditions

Based on transcriptomic analyses, the transcriptional variations of AF cluster genes were listed in Table 1. In comparison of a_w 0.90 vs. 0.95, 24 of 34 AF biosynthetic genes were significantly up-regulated. The 25 genes of the AF cluster were apparently increased in a_w 0.90 than a_w 0.99, and 15 AF biosynthesis genes were significantly up-regulated in a_w 0.95 than a_w 0.99. Among these genes, *aflV*, *aflO*, *aflI*, *aflLa*, and *aflL* showed the most obviously increased in a_w 0.90, but the expression of initial steps genes, *aflA* and *aflB*, were not increased in comparisons of a_w 0.90 vs. 0.95 and a_w 0.95 vs. 0.99. The expressions in different a_w treatments of the pathway-specific regulators, AflR and AflS, showed up-regulations, but were not significantly changed in a_w 0.90 vs. 0.95 and a_w 0.95 vs. 0.99. All these results suggested that transcriptional expressions of the AF cluster genes could be affected by different a_w levels.

Table 1. Comparisons of AF biosynthesis cluster genes in different a_w by transcriptome analysis.

Gene_ID (AFLA_)	Gene	Gene Function	Log$_2$ (90/95)	Log$_2$ (90/99)	Log$_2$ (95/99)
139100	*aflYe*	Ser-Thr protein phosphatase family protein	−0.45	−1.21	−0.78
139110	*aflYd*	sugar regulator	−0.86	−0.33	0.50
139120	*aflYc*	glucosidase	−0.42	−0.59	−0.19
139130	*aflYb*	putative hexose transporter	−0.13	−0.59	−0.48
139140	*aflYa*	NADH oxidase	3.94 *	4.05 *	0.09
139150	*aflY*	hypothetical protein	4.96 *	5.20 *	0.24
139160	*aflX*	monooxygenase	4.58 *	5.92 *	1.33
139170	*aflW*	monooxygenase	4.42 *	6.23 *	1.81 *
139180	*aflV*	cytochrome P450 monooxygenase	5.33 *	12.53 *	7.18 *
139190	*aflK*	VERB synthase	4.79 *	11.23 *	6.43 *
139200	*aflQ*	cytochrome P450 monooxigenase	5.14 *	11.79 *	6.65 *
139210	*aflP*	O-methyltransferase A	5.05 *	11.05 *	5.99 *
139220	*aflO*	O-methyltransferase B	5.03 *	12.05 *	10.83 *
139230	*aflI*	cytochrome P450 monooxigenase	6.21 *	13.05 *	6.95 *
139240	*aflLa*	hypothetical protein	5.40 *	14.05 *	8.11 *
139250	*aflL*	P450 monooxygenase	4.73 *	13.77 *	9.03 *
139260	*aflG*	cytochrome P450 monooxygenase	4.22 *	6.17 *	1.94 *
139270	*aflNa*	hypothetical protein	0.83	1.32	0.48
139280	*aflN*	monooxygenase	4.05*	7.46 *	3.39 *
139290	*aflMa*	hypothetical protein	4.30 *	9.85 *	5.53 *
139300	*aflM*	ketoreductase	4.53 *	12.29 *	7.74 *
139310	*aflE*	NOR reductase	4.34 *	7.97 *	3.63 *
139320	*aflJ*	esterase	4.06 *	6.95 *	2.89 *
139330	*aflH*	short chain alcohol dehydrogenase	3.64 *	5.06 *	1.41
139340	*aflS*	pathway regulator	0.54	3.51 *	0.96
139360	*aflR*	transcription activator	0.43	1.82 *	1.37
139370	*aflB*	fatty acid synthase beta subunit	1.22	2.59 *	1.36
139380	*aflA*	fatty acid synthase alpha subunit	1.73	2.06 *	0.31
139390	*aflD*	reductase	3.35 *	3.73 *	0.37
139400	*aflCa*	hypothetical protein	4.19 *	4.46 *	0.26
139410	*aflC*	polyketide synthase	2.85 *	2.73 *	−0.14
139420	*aflT*	transmembrane protein	−0.10	0.22	0.31
139430	*aflU*	P450 monooxygenase	−0.83	0.15	0.96
139440	*aflF*	dehydrogenase	−0.61	−0.16	0.44

Transcriptome analyses were performed in three biological replicates. Data were calculated with read counts. The values 90/95, 90/99, and 95/99 represented the comparisons of a_w 0.90 vs. 0.95, a_w 0.90 vs. 0.99, and a_w 0.95 vs. 0.99, respectively. Significances were marked as * with $p$adj < 0.05 and log$_2$ratio ≥ 1 or ≤ 1.

2.4. Varying Expressions of Diverse Regulator-Associated AF Biosynthesis in Different a_w Conditions

The expression changes of AF biosynthesis-related regulators were listed in Table S1. The majority regulators' expressions, such as the velvet complex genes, the MAPK pathway genes, and the GPCRs genes, were not significantly different in diverse a_w conditions. However, the bZIP TF, AtfB, was obviously changed at different a_w conditions, and the *atfB* levels showed to be significantly up-regulated in comparisons of a_w 0.90 vs. 0.99 and a_w 0.95 vs. 0.99 (Table S1). The other AF production-related TFs were not noticed any differently at different a_w (Table S1). The transcriptional expressions of the oxylipin genes *ppoB* were significantly up-regulated at lower a_w, while *ppoA* and *ppoC* showed similar levels in different a_w comparisons (Table S1). The calcium-binding protein caleosin gene, *AfPXG*, and the cAMP-dependent protein kinase gene, *pkaC*, were not apparently changed in a_w 0.90 vs. 0.95, whereas they showed significantly increased levels in a_w 0.90 vs. 0.99 and a_w 0.95 vs. 0.99 (Table S1). Concerning SakA, homologous with HogA in *Saccharomyces cerevisiae*, its transcriptional expressions were down-regulated at the lower a_w, but significantly changed only in comparison of a_w 0.90 vs. 0.99 (Table S1).

2.5. Different Expression of the Genes Controlling Conidia Production in Different Water Activities

The transcriptional expressions of several conidia developmental and regulatory genes were also analyzed in transcriptome analyses (Table 2). Six conidial development proteins, including conidiation-specific family protein (AFLA_044790), conidiation proteins Con6 and Con10, conidial hydrophobin RodA and RodB, and conidial pigment biosynthesis oxidase Arb2, showed significantly up-regulated transcription in the lower a_w conditions (Table 2). However, conidial-pigment-biosynthesis-related gene *arp1* and conidiophore-development-related gene *hymA* showed no difference at different a_w (Table 2). Several pieces of research reported that the velvet complex and the developmental signal biosynthesis protein FluG could affect the conidia production. However, *veA*, *laeA*, *velB*, and *fluG* showed similar expression in diverse a_w (Table 2). The transcriptional expressions of the developmental regulator FlbA and the conidiation-related TFs, FlbC and StuA, were also not significantly different at a_w 0.90, 0.95, and 0.99 conditions (Table 2). However, the C_2H_2 type conidia developmental TF gene *brlA* and the developmental regulator genes, *vosA* and *wetA*, showed to be significantly more up-regulated at a_w 0.90 than in a_w 0.95 and 0.99 (Table 2). Taken together, the expressions of conidia developmental proteins and their regulators could be affected by different a_w conditions.

Table 2. Comparisons of conidia-development-related genes in different a_w by transcriptome analysis.

Gene_ID (AFLA_)	Gene Annotation	Log2 (90/95)	Log$_2$ (90/99)	Log$_2$ (95/99)
044790	conidiation-specific family protein	0.42	3.54 *	3.11 *
044800	conidiation protein Con6, putative	3.18 *	8.32 *	5.13 *
083110	conidiation-specific protein (Con10), putative	2.78 *	6.32 *	3.54 *
098380	conidial hydrophobin RodA/RolA	6.49 *	8.68 *	2.18 *
014260	conidial hydrophobin RodB/HypB	3.21 *	3.10 *	−0.13
006180	conidial pigment biosynthesis oxidase Arb2/brown2	5.76 *	6.39 *	0.61
016140	conidial pigment biosynthesis scytalone dehydratase Arp1	−1.57	−1.47	−0.08
079710	conidiophore development protein HymA	−0.01	0.88	0.87
082850	C2H2 type conidiation transcription factor BrlA	3.62 *	5.90 *	2.27 *
029620	transcription factor AbaA	4.19 *	2.52 *	−1.69
134030	developmental regulator FlbA	−0.11	−1.10	−1.01
137320	C2H2 conidiation transcription factor FlbC	−1.10	1.12	1.21
080170	MYB family conidiophore development protein FlbD	−0.60	−0.87	0.28
026900	developmental regulator VosA	2.45 *	1.42 *	−1.05
046990	APSES transcription factor StuA	0.24	1.07	0.81
052030	developmental regulatory protein WetA	2.10 *	2.60 *	0.48
101920	extracellular developmental signal biosynthesis protein FluG	0.06	0.40	0.32

Transcriptome analyses were performed in three biological replicates. Data were calculated with read counts. The values of 90/95, 90/99, and 95/99 represented the comparisons of a_w 0.90 vs. 0.95, a_w 0.90 vs. 0.99, and a_w 0.95 vs. 0.99, respectively. Significances were marked as * with *p*adj < 0.05 and log$_2$ratio ≥ 1 or ≤1.

2.6. The Effects of Diverse Water Activities on Transcription Factors

The TFs' expressions in different a_w were additionally analyzed in this study. In a total of 271 TFs (annotated in this transcriptome data), 29 transcriptional factors showed significant variations in the comparison of a_w 0.90 vs. 0.99 (Table 3). Among them, 20 genes were significantly up-regulated at a_w 0.90, while the other nine genes were significantly down-regulated. With the exception of the two mentioned TFs, BrlA and AtfB, the TFs, including LeuB, RosA, NosA, AbaA, and MeaB, were also significantly increased at a_w 0.90 compared to a_w 0.99. In the comparison of a_w 0.90 vs. 0.95, the expressions of TF genes, AFLA_029620 (*abaA*), AFLA_040300, AFLA_082850 (*brlA*), and Novel 00457 were up-regulated at a_w 0.90. In the comparison of a_w 0.95 vs. 0.99, only *nosA*, *atfB*, and *brlA* levels were increased. So, several TFs genes were affected by a_w conditions, and further regulated the transcriptions of downstream genes.

Table 3. Comparisons of different TFs in different a_w by transcriptome analysis.

Gene ID (AFLA_)	Gene Description	log2 (90/95)	log2 (90/99)	log2 (95/99)
013240	C6 transcription factor, putative	−2.41	−2.10 *	0.30
015790	C6 transcription factor (Leu3), putative	0.19	1.96 *	1.74
021930	C6 transcription factor RosA	0.53	1.74 *	1.19
023040	C6 transcription factor, putative	−3.02	−4.27 *	−1.25
025720	C6 transcription factor NosA	2.46	2.46 *	2.21 *
029620	transcription factor AbaA	4.19 *	2.52 *	−1.69
030580	C2H2 transcription factor PacC, putative	−0.50	−2.02 *	−1.53
031790	bZIP transcription factor (MeaB), putative	−0.56	−1.80 *	−1.26
033480	C6 transcription factor, putative	1.02	1.85 *	0.81
035590	C6 transcription factor, putative	−0.16	2.75 *	2.25
040300	C6 transcription factor, putative	2.36 *	2.75	0.37
051900	zinc knuckle transcription factor (CnjB), putative	0.48	2.73 *	2.23
056780	C6 transcription factor, putative	−0.84	−2.27 *	−1.44
059510	fungal specific transcription factor, putative	−0.95	−1.76 *	−0.84
070970	C6 transcription factor, putative	0.60	1.61 *	1.00
074200	C6 transcription factor, putative	−0.76	−1.90 *	−1.16
076320	C6 transcription factor, putative	1.24	2.61 *	1.35
078500	bZIP transcription factor, putative	0.92	2.65 *	1.72
082850	C2H2 type conidiation transcription factor BrlA	3.62 *	5.90 *	2.27 *
083460	C6 transcription factor RosA-like, putative	−1.64	−1.91 *	−0.28
083560	C6 transcription factor, putative	0.72	2.01 *	1.28
084720	C6 transcription factor, putative	0.68	2.56 *	1.87
085880	BTB domain transcription factor, putative	1.14	1.42 *	0.27
087810	bZIP transcription factor, putative	0.51	2.69 *	2.17
094010	bZIP transcription factor (*Atf21*), putative	1.06	3.69 *	2.60 *
095090	C6 transcription factor, putative	1.87	5.79 *	3.90
109220	C6 transcription factor, putative	0.77	1.95 *	1.16
Novel00457	fungal specific transcription factor [*Aspergillus oryzae* RIB40]	1.72 *	2.25 *	−0.52
Novel00611	transcription factor [*Aspergillus oryzae* RIB40]	−1.08	−3.22 *	−2.16

Transcriptome analyses were performed in three biological replicates. Data were calculated with read counts. The values of 90/95, 90/99, and 95/99 represented the comparisons of a_w 0.90 vs. 0.95, a_w 0.90 vs. 0.99, and a_w 0.95 vs. 0.99, respectively. Significances were marked as * with padj < 0.05 and $\log_2$ratio $\geq$ 1 or $\leq$1.

2.7. RT-qRCR Analyses of Genes Expressions Involved in AF Biosynthesis and Conidia Development

RT-qPCR was performed for confirming the transcriptome results. Similar with transcriptome data, *aflA* and *aflC* were up-regulated at a_w 0.90 compared with a_w 0.95 and 0.99, and *aflK*, *aflO*, and *aflV* were more drastically increased. Additionally, *aflO* in comparison to a_w, 0.90 vs. 0.99 showed the biggest difference with 4.04-$\log_2$FoldChange. The *aflR* was only significantly changed in a_w 0.90 vs. 0.99, while *aflS* levels were increased at a_w 0.90 and 0.95 compared to a_w 0.99 (Figure 4A). The transcripts of *atfB*, *ppoB*, and

AfPXG were significantly up-regulated under the lower a_w conditions, but the expressions of *veA* and *atfA* were not significantly changed (Figure 4A). The conidia developmental genes, *con6*, *con10*, *rodA*, and *rodB*, were significantly up-regulated at a_w 0.90 compared with a_w 0.95 and 0.99. The conidial regulators, *brlA*, *abaA*, and *wetA* were also obviously increased at a_w 0.90, but the other two regulators, *flbA* and *stuA*, had no obvious variations (Figure 4B). In order to verify our results, we also investigated these genes' expressions in other *Aspergillus* strains at different a_w conditions. In *A. flavus* CA14, all AF cluster genes' expressions were similar with *A. flavus* NRRL3357, but with the exception of *atfB*, the expression of *atfA* was also up-regulated in a_w 0.90 compared than a_w 0.99 (Figure S1). In *A. flavus* ACCC32656, both atfA and atfB were increased in the lower a_w conditions, but the *aflA* and *aflC* were not significantly changed (Figure S1). For the conidiation, the conidial genes' expressions were similar in different strains, while the *wetA* in ACC32656 were not significantly varied in diverse a_w conditions.

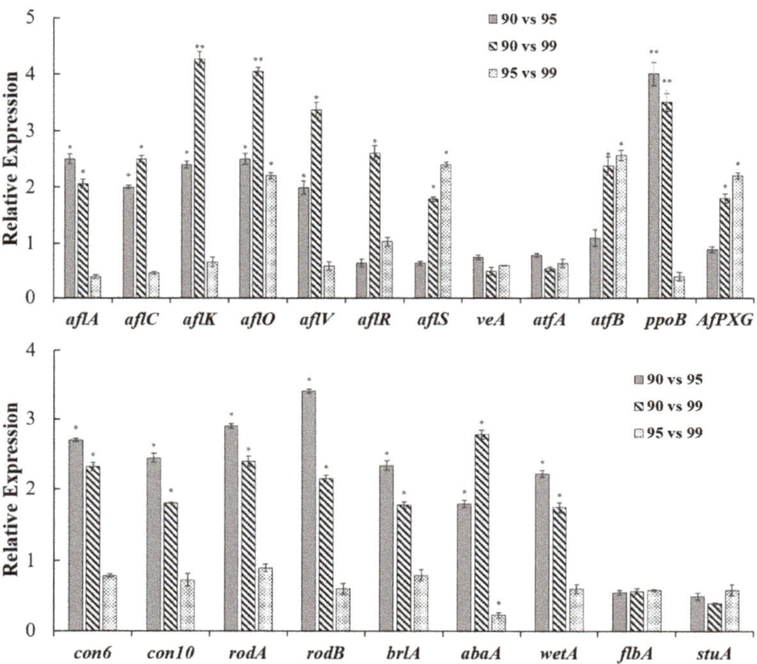

Figure 4. Transcriptional expression analyses of diverse genes by RT-qPCR. The RT-qPCR analysis of (**A**) AF biosynthesis-related genes and (**B**) conidia developmental genes in different a_w conditions. The different a_w comparisons were showed as diverse bars. Three independent biological replicates were performed in each condition, and data were presented as means $\pm$ SD. *t* tests were applied for significance analyses with * $p < 0.05$ and ** $p < 0.01$.

3. Discussion

In this paper, the a_w 0.90 of peanuts showed the maximum AFB$_1$ production after 10 days cultivation (Figure 1D). Abdel-Hadi et al. found that *A. flavus* in peanuts would produce the maximum amounts of AFB$_1$ at a_w 0.90–0.95 after 3 weeks storage [13]. Liu et al. indicated that AFB$_1$ levels were obviously increased in a_w 0.95, followed by a_w 0.90, but were suppressed in a_w 0.99 [6]. The relatively low peanut a_w could be suitable for AF production, and a_w 0.99 could not be a proper condition for AF biosynthesis. We believed that the condition of a_w 0.99 could be a stress signal for *A. flavus*. However, in other studies, the results could be opposite. Zhang et al. found that *A. flavus* produced more AFB$_1$ in a_w 0.99 than at a_w 0.93 in YES medium, and Medina et al. noticed that AFB$_1$

levels of maize were lower in a_w 0.91 than 0.99 [8,9]. It seems like the suitable a_w levels could be varied depending on diverse substrates. Different temperatures also influence the optimum a_w for AF biosynthesis. The optimal a_w for AF biosynthesis was 0.92 upon 28 °C, while it increased to 0.96 at the lower temperature [14]. Further, the effect of a_w on AF production was apparently modulated by the stages of cultivation, maturity, and storage [15]. Strain-specificity is another important reason for different AF productions, such as *A. flavus* CA14 showing the highest AF production in a_w 0.95 [6], but *A. flavus* NRRL3357 showing the most AF levels in a_w 0.90. Taken all this, it is concluded that a_w is a crucial factor for AF biosynthesis, and the effect of a_w on AF production is dependent on other environmental factors, such as temperature, substrates, pH, cultivation time, and different strains. Because of the diverse experiment conditions, it is hard to get a consistent result. So, in this study, we focused our research on the regulatory mechanism of a_w on AF biosynthesis.

AF cluster gene expressions are directly related to AF biosynthesis. There are some studies reporting the variations of AF gene expression in different a_w. Most AF genes had higher expression levels at lower a_w [6], and *aflD* showed higher expression at aw 0.90 [13]. In this study, we examined the transcriptional expressions of AF cluster genes by RNA-seq and RT-qPCR analyses (Table 1 and Figure 4A). The majority of genes (27/34) in AF clusters were significantly up-regulated at the relatively lower a_w (90 and 95) (Table 1). These results differed from previous reports [16,17], but were similar with Liu et al. [6]. The AF biosynthetic initial-genes, *aflA*, *aflB*, *aflC*, and *aflD*, showed slight or moderate variations at different a_w (Table 1 and Figure 4A). Abdel-Hadi et al. suggested the initial step gene *aflD* was a good indicator of AFB_1 production [13]. However, in our study, *aflD* expressions in a_w 0.95 vs. 0.99 were not significantly different, and were mildly changed in a_w 0.95 vs. 0.99 and a_w 0.95 vs. 0.99 (Table 1). Ehrlich suggested that the later stages of AFB_1 biosynthesis were more critical than the beginning stages [18]. In our study, the AF cluster genes in medium or later stages, such as *aflI*, *aflO*, *aflP*, *aflQ*, *aflK*, and *aflV*, showed more drastic variations in different a_w conditions. All the above information indicated that AF biosynthesis was influenced by different a_w, especially the biosynthetic process from norsolorinic acid (NOR) to O-methylsterigmatocystin (OMST).

Transcriptions of AF biosynthetic genes are mainly regulated by the cluster-specific regulators, AflR and AflS, which directly bind to the promoter region of AF cluster genes [19]. In our research, *aflR* and *aflS* levels in *A. flavus* NRRL3357 and ACCC32656 showed the moderate increases at a_w 0.90 vs. 0.95, while no significant variations of *aflR* and *aflS* were noticed in the other two a_w comparisons (Table 1 and Figure 4A). However, in *A. flavus* CA14, *aflR* and *aflS* were increased in a_w 0.90 compared with a_w 0.99 (Figure S1), suggesting the AF cluster-specific regulators might be affected in different strains upon the diverse a_w. There are also many studies that found that the ratio of *aflS*/*aflR* should have the closer correlation with AF productions [9,11,17]. However, in this research, the ratios of *aflS*/*aflR* were still similar in different a_w treatments. So, the transcriptional changes of AF structural genes could not be only caused by the changes of *aflR* and *aflS*, but other regulators could play more important roles.

Furthermore, there are some papers reporting that the expressions of AF cluster genes were influenced by different environmental factors. However, few of them focused on how a_w affected AF genes' expression, and what the critical regulator response to a_w is. In this study, to deeply investigate the reasons of AF gene variations in different a_w, the comprehensive transcriptomic analysis was performed, and the oxidation-stress-related TFs, AtfA, AtfB, AP-1, MsnA, MtfA, and SrrA, were also examined, which could control the AF cluster gene transcriptions by directly binding [12,20,21]. However, in this study, the above TF genes, with the exception of AtfB, showed similar transcriptional expressions at different a_w (Table S2 and Figure 4A). The *atfB* expression was significantly different in different a_w conditions (Table S2 and Figure 4A), suggesting AtfB should be a key responder of a_w conditions. AtfB, as a member of CREB family protein, could recognize the CRE binding sites (5′-TG/TACGTC/AA-3′), and start the target gene transcript [12].

In *A. parasiticus*, in the upstream noncoding regions of *aflB*, *aflD*, *aflM*, *aflO*, and *aflR*, were found the CRE sites, which could be directly bound by AtfB [22]. So, their transcriptional expressions were positively correlated with *atfB* expression. Suppression of AtfB could significantly reduce the AF genes' mRNA levels and the AF production [23]. Similarly, in this study, significantly more down-regulation of *atfB* was found at a_w 0.95 and 0.99 than a_w 0.90; subsequently, most AF genes and AF productions also were decreased at the higher a_w conditions. In recent research, AtfB was suppressed by methyl jasmonate, and subsequently, down-regulated AF gene expressions [24]. So, AtfB is a critical regulator for sensing and response to environmental changes, and then could modulate downstream genes, such as AF cluster genes in *A. flavus*. Additionally, we also tested the *atfB* expression in other *Aspergillus* strains, of which the *atfB* in *A. flavus* CA14 and *A. flavus* ACCC 32656 were significantly up-regulated in a_w 0.90 (Figure S1). All these results that confirmed the differential expression of *atfB* in different a_w treatments might play a vital role in the changes of AF genes' expressions and AF production.

The environmental signals could be sensed by the membrane protein, transferred by the phosphorylation signal, and responded to by TFs. For example, the oxidation stresses up-regulate SAPK/MAPK signaling cascade, and then activate AtfB for binding to the target promoters [12]. In this study, *sakA2* (AFLA_099500), a kinase of MAPK pathway, is slightly down-regulated in a_w 0.90 vs. 0.99, suggesting it could be affected by different a_w conditions (Table S2). However, we did not find other differential transcriptional expressions of MAPK genes in different a_w conditions (Table S2). It could be explained that the MAPK cascade transmits the signal by phosphorylation, and the effect of different a_w on MAPK genes could be at a post-transcriptional level. *pkaC*, an encoding cAMP-dependent protein kinase catalytic subunit, was significantly more down-regulated at a_w 0.99 than at a_w 0.90 and 0.95 (Table S2). The cAMP/PKA pathway can also regulate AF biosynthesis partly through AtfB [23,25], and AtfB responds to carbon sources and oxidative stress through the cAMP pathway [22]. It is a reasonable hypothesis that *pkaC* levels are modulated at different a_w levels, and then affect AtfB expression by the cAMP signaling pathway.

In previous studies, the conidia production and conidia germination of *Aspergillus* strains and *Penicillium* strains were significantly affected by different a_w levels [26,27]. We also noticed that the apparently decreased conidia production at a_w 0.99 in peanuts (Figure 1C), and transcriptions of conidial genes, were also significantly decreased at a_w 0.99 (Table 2 and Figure 4B). The *con6* and *con10*, as the representatives of conidiation genes, are conserved in filamentous fungi and preferentially expressed during the conidia development [28]. In *A. nidulans*, *conF* (homologous with *con6*) and *conJ* (homologous with *con10*) were increased with light exposure [29]. Similarly, their expressions at different a_w were obviously changed (Table 2 and Figure 4B), suggesting that *con* genes may be affected by diverse environmental factors. RodA and RodB, as the hydrophobin proteins, help conidia dispersion and attachment [30], and their transcriptions were also increased at the lower a_w (Table 2 and Figure 4B). It is also noticed that the conidial pigment-related gene, *arb2*, was significantly down-regulated in a_w 0.99 (Table 2). It could partly explain why the green color was faded in the higher a_w conditions (Figure 1A,B).

Conidia-relevant regulators, BrlA, AbaA, VosA, and WetA, were also significantly increased in a_w 0.90, and decreased in a_w 0.99 (Table 2 and Figure 4B). BrlA, as the C_2H_2 zinc finger TF, governs the *wetA* and *abaA* expressions, and positively regulates conidia production [31]. The transcript of *abaA* is promoted by BrlA in the middle stages of conidia development, and involved in the differentiation and functionality of phialides [32]. Lack of AbaA leads to the decreased and aberrant conidia production [33]. *wetA* is regulated by AbaA during the late phase of conidia development, and plays a role in the conidial wall component biosynthesis [34]. Based on previous research, deletion of any of the three genes could interfere with the conidial genes' expression and conidial development. In this study, few conidia were produced at a_w 0.99, and conidiation-related genes were also significantly down-regulated. It is supposed that a_w might regulate conidia development through the

BrlA-AbaA-WetA cascade. In addition, the *brlA* expressions of both *A. flavus* CA14 and *A. flavus* ACCC 32656 were significantly up-regulated in lower a_w, but *wetA* in *A. flavus* ACCC 32656 showed no change in different treatments (Figure S1), suggesting that other regulators might be affected by *wetA* expression in *A. flavus* ACCC 32656. VosA is also a multifunctional regulator, interacting with VelB and VelC, and controls conidial trehalose amount and conidial germination in *A. fumigatus* [35,36]. We also noticed significantly increased *vosA* expression at a_w 0.90, but no obvious difference in other velvet complex genes (*veA*, *velB*, and *velC*). The other conidial regulators, FluG, FlbA, FlbC, FlbD, and StuA, [37], were not significantly regulated at diverse a_w (Table 2 and Figure 4B). Furthermore, AtfB was positively relevant with conidia production in *A. oryzae* [38], suggesting AtfB could also be a conidial regulator. In this study, AF production, conidia development, as well as *atfB* expression, showed similar changes in diverse a_w conditions, suggesting that AtfB might be a critical linker of fungal development and secondary metabolism.

Taken together, the deduced regulatory pathway of different a_w effects on AF biosynthesis and conidia development were presented in Figure 5. As Figure 5 shows, different a_w signals affect cellular signaling pathways by modulating the expressions of GPCRs and oxylipins genes; then, several TFs, especially AtfB, are activated by SAPK/MAPK and cAMP/PKA pathways through the multistep phosphorelay systems [12,25]; the up-regulated AtfB can directly bind to the promoter regions of AflR, AflS, and AF biosynthetic genes, and subsequently enhance AF production [12,22]. BrlA, as the central regulator of conidiation, could be up-regulated by a_w 0.90, then motivate AbaA and WetA, and subsequently regulate conidial gene expressions. There are still a lot ambiguous specific regulations in this pathway, and more research is needed to clarify the regulatory mechanism of a_w on AF production and *A. flavus* development.

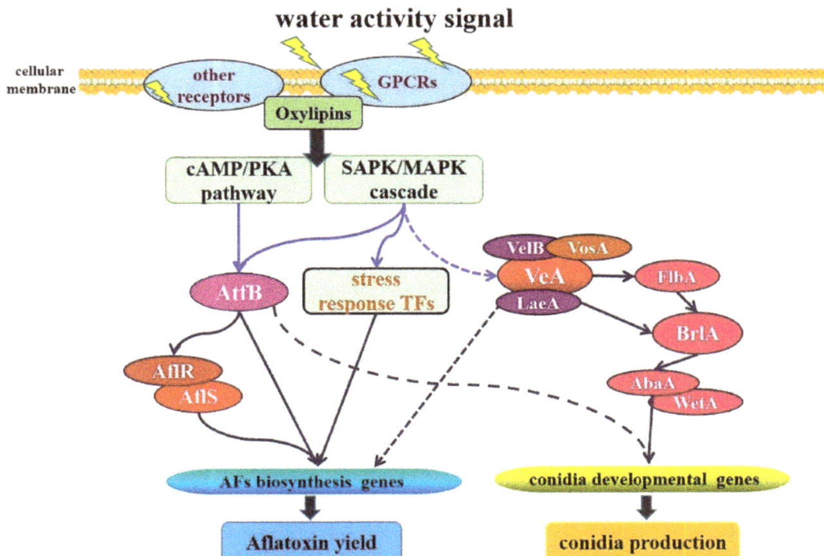

Figure 5. Hypothetical regulatory mechanism of a_w on AF biosynthesis and conidia development. The confirmed regulatory pathway and deduced regulatory pathway were presented as solid lines and dashed lines, respectively. TFs stands for transcription factors.

For better revealing of the transcriptional regulations in different a_w, we also detected the expressions of diverse TFs. Among 271 annotated TFs, 29 TFs were significantly changed, including *leuB*, *rosA*, *nosA*, *abaA*, *meaB*, *brlA*, *atfB*, etc. (Table 3). NosA and RosA, as the $Zn(II)_6Cys_6$ class activators, are homologous with Pro1 in *Sordaria macrospora*, and regulate sexual development in *Aspergillus* [39]. However, RosA represses sexual development in the early stage, while NosA is necessary for primordium maturation [40]. The significant increase of *nosA* and *rosA* was observed at a_w 0.90 vs. 0.99, suggesting that sexual development of A. *flavus* may be affected by diverse a_w levels. MeaB as the methylammonium-resistant protein, is involved in nitrogen metabolite repression, and positively regulates sterigmatocystin production in A. *nidulans* [41]. However, in A. *flavus*, *meaB* was up-regulated at the higher a_w condition, and was negatively relevant with AF production (Table 3). LeuB/Leu3 participates in branched-chain amino acids biosynthesis, *gdhA* expression, as well as nitrogen metabolism, and physically interacts with AreA [42,43]. Moreover, by KEGG analysis, DEGs were obviously enriched in nitrogen metabolite (Figure 3). All information indicated that nitrogen metabolite of A. *flavus* in peanuts was also affected by diverse a_w levels.

4. Conclusions

In this study, A. *flavus* strain NRRL3357 was inoculated in peanuts with diverse a_w (0.90, 0.95, and 0.99). The changes of AFB$_1$ yield and conidia production showed the highest level in a_w 0.90, followed by a_w 0.95, and the minimal level in a_w 0.99. Based on transcriptome data and RT-qPCR analyses, we noticed that (1) most of the AF biosynthesis genes were more up-regulated in a_w 0.90 than a_w 0.95 and 0.99; (2) the initial-step AF genes were slightly or moderately changed, while the middle- or later-step genes showed drastic responses to different a_w conditions; (3) several kinases, membrane proteins, and TFs were affected by different a_w, and AtfB could be the central TF for regulating the transcriptional expressions of downstream genes, especially AF structural genes; (4) conidia development genes and the conidial regulator genes were up-regulated in a_w 0.90; (5) sexual-development-relevant TFs, NosA and RosA, and nitrogen-metabolite-relevant TFs, MeaB and LeuB, were significantly changed at diverse a_w.

5. Materials and Methods

5.1. Fungal Strain and Conidia Suspension Preparation

A. *flavus* NRRL3357 and ACCC32656 were kindly provided by Professor Wenbing Yin (Institute of Microbiology, Chinese Academy of Sciences, Beijing, China). A. *flavus* CA14 was kindly provided by Professor Shihua Wang (Fujian Agriculture and Forestry University, Fujian, China). The strains were stored at $-80\ °C$ and re-cultivated on PDA medium (200 g potato, 20 g glucose, and 20 g agar in 1 L distilled water) at $28\ °C$ in the dark. Conidia were harvested from PDA plates after 7 days inoculation by 0.01% Tween 20, and the suspension concentration was counted by hemocytometer, and was adjusted as 10^7 conidia/mL.

5.2. Adjustment of Peanut Water Activities and Inoculation of A. flavus Conidia Suspension

The method of a_w adjusting was followed as that by Liu et al. with some modifications. The a_w levels were detected by the Aqualab 4TE (Decagon Devices, Pullman, WA, USA), and the a_w curve of peanuts was performed in pre-experiment for accurately defining the amount of water added into the peanuts [6]. For adjusting the specific a_w, 100 g of peanuts were put into zip-lock bags, irradiated with UV light for 2 h, and then the determined amount of water was added to them to obtain targeted a_w levels (a_w 0.90, 0.95, and 0.99). All treatments were placed in $4\ °C$ overnight for the stable a_w levels.

Then these treated peanuts were transferred into the 500 mL sterile flasks, and incubated in 10 mL of the 10^7 conidia/mL conidia suspension. Fungi in different a_w levels were cultivated at $28\ °C$ for 10 days in the polyethylene boxes, which contained the glycerol-water solution for maintaining the relatively constant humidity. Peanut kernels without inoculating conidia suspension were prepared as a negative control. Each flask was shaken once a day. Three biological replicates were performed for all treatments.

5.3. Conidia Assessment and AFB1 Detection

After 10 days cultivation, 25 g of inoculated peanuts with different a_w were added 100 mL sterilized H_2O, fiercely shaken for 30 min, filtered with non-woven fabric, and conidia of the solution was counted by a hemocytometer.

AFB_1 concentration was detected by HPLC analysis. An amount of 25 g of peanut samples were finely grounded, 125 mL 70% methanol water and 5 g NaCl were added, and fiercely vibrated for 30 min. AFB_1 extractions was purified by ToxinFast immunoaffinity columns as per the manufacturer's instructions (Huaan Magnech Biotech, Beijing, China), and were examined by an Agilent 1220 Infinity II HPLC system coupled with a fluorescence detector and a post-column derivation system (Huaan Magnech Biotech, Beijing, China). The excitation wavelength was 360 nm, and the emission wavelength was 430 nm. The HPLC system was matched with the Agilent TC-C18 column (250 mm × 4.6 mm, 5 μm particle size, Agilent). An amount of 20 μL AFB_1 samples were injected each time, 70% methanol solution was the mobile phase, and the retention time was about 5.7 min. AFB_1 standards were purchased from Sigma-Aldrich (St. Louis, MO, USA).

5.4. Total RNA Extraction

RNA samples for transcriptome analysis and RT-qPCR were performed three times by replications. Mycelia were harvested from the inoculated peanuts' seed coats after 10 days cultivation. An amount of 1 g samples (the mixture of peanut seed coat and *A. flavus* mycelia) were grounded to powder after treated by liquid nitrogen, then 600 μL lysis buffer was added, and then the RNA was extracted as per the manufacturer's instructions (Aidlab, Beijing, China). Genomic DNA was removed by DNase I (Takara, Dalian, China), and RNA quality was evaluated by NanoDrop 2000 spectrophotometer (Thermo Fisher, Waltham, MA, USA) and Agilent 2100 Bioanalyzer (Agilent, Santa Clara, CA, USA).

5.5. RNA Sequencing and Transcriptome Processing

The mRNA was sequenced by Novogene (Beijing, China). Briefly, mRNA was purified from total RNA with oligo-dT magnetic beads. The non-strand-specific libraries were constructed by NEB Next Ultra^{TM} RNA Library Prep Kit for Illumina (NEB, USA), and sequenced by the Illumina Hiseq 4000 platform (Illumina Inc., San Diego, CA, USA). Clean reads were harvested by removing the low-quality reads and adaptor, and then mapped to the reference genome (BioProject: PRJNA13284) with HISAT 1.31 [44]. The read counts were used to assess genes' transcriptions [45]. The differentially expressed genes (DEGs) were evaluated with $p_{adj} \leq 0.05$ and $\log_2$ratio ≥ 1 or ≤ 1. The Gene Ontology (GO) functional analysis and Kyoto Encyclopedia of Genes and Genomes (KEGG) pathway analysis of DEGs were performed with the FungiFun and KAAS, respectively [46,47].

5.6. RT-qPCR Analysis

Total RNA was used for reverse transcription, and cDNA synthesis was with a two-step cDNA synthesis kit (TaKaRa, Dalian, China). The Analytic Jena Q-tower system (Analytik-Jena, Jena, Germany) was used for qPCR assays with the 20 μL reaction system, including 5 μL cDNA product, 0.5 μL of each primer, and 10 μL SYBR Green mix (TaKaRa, Dalian, China). All primers are listed in Table S2. The qPCR program was settled as before, which is one cycle of 3 min at 95 °C followed by 40 cycles of 10 s at 95 °C and 40 s at 65 °C, and the melting curve was analyzed from 60 °C to 90 °C with 0.5 °C incremental increases. The internal reference was used with *actin*. The transcriptional expression was based on the CT value, and the differences were calculated with the $2^{-\Delta\Delta CT}$ method.

5.7. Statistical Analysis

Three biological replicates were performed for all experiments. The means with standard deviations represented the results. AFB_1 yields and conidia productions in different treatments were calculated with one-way analysis of variance (ANOVA) by SPSS

18.0, and statistical differences were evaluated by Tukey's test with $p < 0.05$. Student's t test was applied in RT-qPCR with * $p < 0.05$ and ** $p < 0.01$.

Supplementary Materials: The following are available online at https://www.mdpi.com/article/10.3390/toxins13060431/s1, Figure S1: Transcriptional expression analyses of diverse genes by RT-qPCR, Table S1: Comparisons of several global regulators in different a_w by transcriptome analysis, Table S2: Primers used for qPCR analysis.

Author Contributions: Conceptualization, X.Y.; data curation, Q.Y.; formal analysis, C.N.; funding acquisition, Y.L. and Y.Z.; investigation, X.L., X.M. and Q.Y.; project administration, X.Y.; resources, Q.Y., Y.L., Y.Z. and F.X.; supervision, F.X.; validation, X.M.; writing—original draft, L.M. and X.L.; Writing—review & editing, X.L. All authors have read and agreed to the published version of the manuscript.

Funding: This research was funded by National Natural Science Foundation of China (32001813 and 31972179), Qingdao Science and Technology Benefit the People Demonstration and Guidance Special Project (21-1-4-NY-4-NSH), Key R&D Program of Zhangjiakou (19120002D), and National Agricultural Science and Technology Innovation Program (CAAS-ASTIP-2021-IFST).

Institutional Review Board Statement: Not applicable.

Informed Consent Statement: Not applicable.

Data Availability Statement: All data are provided in the manuscript.

Conflicts of Interest: The authors declare no conflict of interest.

References

1. Zhang, F.; Zhong, H.; Han, X.; Guo, Z.; Yang, W.; Liu, Y.; Yang, K.; Zhuang, Z.; Wang, S. Proteomic profile of *Aspergillus flavus* in response to water activity. *Fungal Biol.* **2015**, *119*, 114–124. [CrossRef]
2. Ren, Y.; Jin, J.; Zheng, M.; Yang, Q.; Xing, F. Ethanol Inhibits Aflatoxin B(1) Biosynthesis in *Aspergillus flavus* by Up-Regulating Oxidative Stress-Related Genes. *Front. Microbiol.* **2019**, *10*, 2946. [CrossRef]
3. Wu, F. Perspective: Time to face the fungal threat. *Nature* **2014**, *516*, S7. [CrossRef] [PubMed]
4. Abdel-Hadi, A.; Schmidt-Heydt, M.; Parra, R.; Geisen, R.; Magan, N. A systems approach to model the relationship between aflatoxin gene cluster expression, environmental factors, growth and toxin production by *Aspergillus flavus*. *J. R. Soc. Interface* **2012**, *9*, 757–767. [CrossRef]
5. Tai, B.; Chang, J.; Liu, Y.; Xing, F. Recent progress of the effect of environmental factors on *Aspergillus flavus* growth and aflatoxins production on foods. *Food Qual. Saf.* **2020**, *4*, 21–28. [CrossRef]
6. Liu, X.; Guan, X.; Xing, F.; Lv, C.; Dai, X.; Liu, Y. Effect of water activity and temperature on the growth of *Aspergillus flavus*, the expression of aflatoxin biosynthetic genes and aflatoxin production in shelled peanuts. *Food Control.* **2017**, *82*, 325–332. [CrossRef]
7. Passamani, F.R.; Hernandes, T.; Lopes, N.A.; Bastos, S.C.; Santiago, W.D.; Cardoso, M.; Batista, L.R. Effect of temperature, water activity, and pH on growth and production of ochratoxin A by *Aspergillus niger* and *Aspergillus carbonarius* from Brazilian grapes. *J. Food Prot.* **2014**, *77*, 1947–1952. [CrossRef] [PubMed]
8. Medina, A.; Gilbert, M.K.; Mack, B.M.; GR, O.B.; Rodríguez, A.; Bhatnagar, D.; Payne, G.; Magan, N. Interactions between water activity and temperature on the *Aspergillus flavus* transcriptome and aflatoxin B(1) production. *Int. J. Food Microbiol.* **2017**, *256*, 36–44. [CrossRef] [PubMed]
9. Zhang, F.; Guo, Z.; Zhong, H.; Wang, S.; Yang, W.; Liu, Y.; Wang, S. RNA-Seq-based transcriptome analysis of aflatoxigenic *Aspergillus flavus* in response to water activity. *Toxins* **2014**, *6*, 3187–3207. [CrossRef] [PubMed]
10. Yu, J. Current understanding on aflatoxin biosynthesis and future perspective in reducing aflatoxin contamination. *Toxins* **2012**, *4*, 1024–1057. [CrossRef]
11. Yu, J.; Fedorova, N.D.; Montalbano, B.G.; Bhatnagar, D.; Cleveland, T.E.; Bennett, J.W.; Nierman, W.C. Tight control of mycotoxin biosynthesis gene expression in *Aspergillus flavus* by temperature as revealed by RNA-Seq. *FEMS Microbiol. Lett.* **2011**, *322*, 145–149. [CrossRef] [PubMed]
12. Hong, S.Y.; Roze, L.V.; Linz, J.E. Oxidative stress-related transcription factors in the regulation of secondary metabolism. *Toxins* **2013**, *5*, 683–702. [CrossRef]
13. Abdel-Hadi, A.; Carter, D.; Magan, N. Temporal monitoring of the nor-1 (aflD) gene of *Aspergillus flavus* in relation to aflatoxin B$_1$ production during storage of peanuts under different water activity levels. *J. Appl. Microbiol.* **2010**, *109*, 1914–1922. [CrossRef] [PubMed]
14. Lv, C.; Jin, J.; Wang, P.; Dai, X.; Liu, Y.; Zheng, M.; Xing, F. Interaction of water activity and temperature on the growth, gene expression and aflatoxin production by *Aspergillus flavus* on paddy and polished rice. *Food Chem.* **2019**, *293*, 472–478. [CrossRef]

15. Peromingo, B.; Rodríguez, A.; Bernáldez, V.; Delgado, J.; Rodríguez, M. Effect of temperature and water activity on growth and aflatoxin production by *Aspergillus flavus* and *Aspergillus parasiticus* on cured meat model systems. *Meat Sci.* **2016**, *122*, 76–83. [CrossRef] [PubMed]

16. Gallo, A.; Solfrizzo, M.; Epifani, F.; Panzarini, G.; Perrone, G. Effect of temperature and water activity on gene expression and aflatoxin biosynthesis in *Aspergillus flavus* on almond medium. *Int. J. Food Microbiol.* **2016**, *217*, 162–169. [CrossRef]

17. Schmidt-Heydt, M.; Abdel-Hadi, A.; Magan, N.; Geisen, R. Complex regulation of the aflatoxin biosynthesis gene cluster of *Aspergillus flavus* in relation to various combinations of water activity and temperature. *Int. J. Food Microbiol.* **2009**, *135*, 231–237. [CrossRef] [PubMed]

18. Ehrlich, K.C. Predicted roles of the uncharacterized clustered genes in aflatoxin biosynthesis. *Toxins* **2009**, *1*, 37–58. [CrossRef] [PubMed]

19. Yu, J.; Chang, P.K.; Ehrlich, K.C.; Cary, J.W.; Bhatnagar, D.; Cleveland, T.E.; Payne, G.A.; Linz, J.E.; Woloshuk, C.P.; Bennett, J.W. Clustered pathway genes in aflatoxin biosynthesis. *Appl. Environ. Microbiol.* **2004**, *70*, 1253–1262. [CrossRef]

20. Caceres, I.; El Khoury, R.; Bailly, S.; Oswald, I.P.; Puel, O.; Bailly, J.D. Piperine inhibits aflatoxin B1 production in *Aspergillus flavus* by modulating fungal oxidative stress response. *Fungal Genet Biol.* **2017**, *107*, 77–85. [CrossRef]

21. Zhuang, Z.; Lohmar, J.M.; Satterlee, T.; Cary, J.W.; Calvo, A.M. The Master Transcription Factor mtfA Governs Aflatoxin Production, Morphological Development and Pathogenicity in the Fungus *Aspergillus flavus*. *Toxins* **2016**, *8*, 29. [CrossRef] [PubMed]

22. Roze, L.V.; Chanda, A.; Wee, J.; Awad, D.; Linz, J.E. Stress-related transcription factor AtfB integrates secondary metabolism with oxidative stress response in aspergilli. *J. Biol. Chem.* **2011**, *286*, 35137–35148. [CrossRef] [PubMed]

23. Wee, J.; Hong, S.Y.; Roze, L.V.; Day, D.M.; Chanda, A.; Linz, J.E. The Fungal bZIP Transcription Factor AtfB Controls Virulence-Associated Processes in *Aspergillus parasiticus*. *Toxins* **2017**, *9*, 287. [CrossRef] [PubMed]

24. Li, X.; Ren, Y.; Jing, J.; Jiang, Y.; Yang, Q.; Luo, S.; Xing, F. The inhibitory mechanism of methyl jasmonate on *Aspergillus flavus* growth and aflatoxin biosynthesis and two novel transcription factors are involved in this action. *Food Res. Int.* **2021**, *140*, 110051. [CrossRef] [PubMed]

25. Roze, L.V.; Miller, M.J.; Rarick, M.; Mahanti, N.; Linz, J.E. A novel cAMP-response element, CRE1, modulates expression of nor-1 in *Aspergillus parasiticus*. *J. Biol. Chem.* **2004**, *279*, 27428–27439. [CrossRef] [PubMed]

26. Long, N.; Vasseur, V.; Coroller, L.; Dantigny, P.; Rigalma, K. Temperature, water activity and pH during conidia production affect the physiological state and germination time of Penicillium species. *Int. J. Food Microbiol.* **2017**, *241*, 151–160. [CrossRef]

27. Pardo, E.; Lagunas, U.; Sanchis, V.; Ramos, A.J.; Marín, S. Influence of water activity and temperature on conidial germination and mycelial growth of ochratoxigenic isolates of *Aspergillus ochraceus* on grape juice synthetic medium. Predictive models. *J. Sci. Food Agric.* **2005**, *85*, 1681–1686. [CrossRef]

28. Olmedo, M.; Ruger-Herreros, C.; Luque, E.M.; Corrochano, L.M. A complex photoreceptor system mediates the regulation by light of the conidiation genes con-10 and con-6 in Neurospora crassa. *Fungal Genet Biol.* **2010**, *47*, 352–363. [CrossRef]

29. Suzuki, S.; Sarikaya Bayram, Ö.; Bayram, Ö.; Braus, G.H. conF and conJ contribute to conidia germination and stress response in the filamentous fungus *Aspergillus nidulans*. *Fungal Genet Biol.* **2013**, *56*, 42–53. [CrossRef]

30. Pedersen, M.H.; Borodina, I.; Moresco, J.L.; Svendsen, W.E.; Frisvad, J.C.; Søndergaard, I. High-yield production of hydrophobins RodA and RodB from *Aspergillus fumigatus* in Pichia pastoris. *Appl. Microbiol. Biotechnol.* **2011**, *90*, 1923–1932. [CrossRef]

31. Twumasi-Boateng, K.; Yu, Y.; Chen, D.; Gravelat, F.N.; Nierman, W.C.; Sheppard, D.C. Transcriptional profiling identifies a role for BrlA in the response to nitrogen depletion and for StuA in the regulation of secondary metabolite clusters in *Aspergillus fumigatus*. *Eukaryot Cell* **2009**, *8*, 104–115. [CrossRef] [PubMed]

32. Sewall, T.C.; Mims, C.W.; Timberlake, W.E. abaA controls phialide differentiation in *Aspergillus nidulans*. *Plant Cell* **1990**, *2*, 731–739. [CrossRef] [PubMed]

33. Andrianopoulos, A.; Timberlake, W.E. The *Aspergillus nidulans* abaA gene encodes a transcriptional activator that acts as a genetic switch to control development. *Mol. Cell. Biol.* **1994**, *14*, 2503–2515. [CrossRef] [PubMed]

34. Marshall, M.A.; Timberlake, W.E. *Aspergillus nidulans* wet A activates spore-specific gene expression. *Mol. Cell. Biol.* **1991**, *11*, 55–62. [CrossRef] [PubMed]

35. Park, H.S.; Nam, T.Y.; Han, K.H.; Kim, S.C.; Yu, J.H. VelC positively controls sexual development in *Aspergillus nidulans*. *PLoS ONE* **2014**, *9*, e89883. [CrossRef]

36. Park, H.S.; Bayram, O.; Braus, G.H.; Kim, S.C.; Yu, J.H. Characterization of the velvet regulators in *Aspergillus fumigatus*. *Mol. Microbiol.* **2012**, *86*, 937–953. [CrossRef]

37. Ni, M.; Yu, J.H. A novel regulator couples sporogenesis and trehalose biogenesis in *Aspergillus nidulans*. *PLoS ONE* **2007**, *2*, e970. [CrossRef]

38. Sakamoto, K.; Arima, T.H.; Iwashita, K.; Yamada, O.; Gomi, K.; Akita, O. *Aspergillus oryzae* atfB encodes a transcription factor required for stress tolerance in conidia. *Fungal Genet Biol.* **2008**, *45*, 922–932. [CrossRef]

39. Vienken, K.; Fischer, R. The Zn(II)2Cys6 putative transcription factor NosA controls fruiting body formation in *Aspergillus nidulans*. *Mol. Microbiol.* **2006**, *61*, 544–554. [CrossRef]

40. Soukup, A.A.; Farnoodian, M.; Berthier, E.; Keller, N.P. NosA, a transcription factor important in *Aspergillus fumigatus* stress and developmental response, rescues the germination defect of a laeA deletion. *Fungal Genet Biol.* **2012**, *49*, 857–865. [CrossRef]

41. Wong, K.H.; Hynes, M.J.; Todd, R.B.; Davis, M.A. Transcriptional control of nmrA by the bZIP transcription factor MeaB reveals a new level of nitrogen regulation in *Aspergillus nidulans*. *Mol. Microbiol.* **2007**, *66*, 534–551. [CrossRef] [PubMed]

42. Downes, D.J.; Davis, M.A.; Kreutzberger, S.D.; Taig, B.L.; Todd, R.B. Regulation of the NADP-glutamate dehydrogenase gene gdhA in *Aspergillus nidulans* by the Zn(II)2Cys6 transcription factor LeuB. *Microbiology* **2013**, *159*, 2467–2480. [CrossRef] [PubMed]

43. Polotnianka, R.; Monahan, B.J.; Hynes, M.J.; Davis, M.A. TamA interacts with LeuB, the homologue of Saccharomyces cerevisiae Leu3p, to regulate gdhA expression in *Aspergillus nidulans*. *Mol. Genet Genom.* **2004**, *272*, 452–459. [CrossRef] [PubMed]

44. Kim, D.; Langmead, B.; Salzberg, S.L. HISAT: A fast spliced aligner with low memory requirements. *Nat. Methods* **2015**, *12*, 357–360. [CrossRef] [PubMed]

45. Trapnell, C.; Williams, B.A.; Pertea, G.; Mortazavi, A.; Kwan, G.; van Baren, M.J.; Salzberg, S.L.; Wold, B.J.; Pachter, L. Transcript assembly and quantification by RNA-Seq reveals unannotated transcripts and isoform switching during cell differentiation. *Nat. Biotechnol.* **2010**, *28*, 511–515. [CrossRef] [PubMed]

46. Priebe, S.; Linde, J.; Albrecht, D.; Guthke, R.; Brakhage, A.A. FungiFun: A web-based application for functional categorization of fungal genes and proteins. *Fungal Genet Biol.* **2011**, *48*, 353–358. [CrossRef] [PubMed]

47. Kanehisa, M.; Araki, M.; Goto, S.; Hattori, M.; Hirakawa, M.; Itoh, M.; Katayama, T.; Kawashima, S.; Okuda, S.; Tokimatsu, T.; et al. KEGG for linking genomes to life and the environment. *Nucleic Acids Res.* **2008**, *36*, D480–D484. [CrossRef] [PubMed]

Article

Toxic Effects of Mycotoxin Fumonisin B1 at Six Different Doses on Female BALB/c Mice

Zhiwei Chen, Fan Zhang, Lin Jiang, Zihan Chen and Hua Sun *

State Key Laboratory of Bioactive Substance and Function of Natural Medicines, Institute of Materia Medica, Chinese Academy of Medical Sciences & Peking Union Medical College, 1# Xiannongtan Street, Xicheng District, Beijing 100050, China; chenzw@imm.ac.cn (Z.C.); AFUN008034@163.com (F.Z.); gillianjiang@imm.ac.cn (L.J.); chenzihan@imm.ac.cn (Z.C.)
* Correspondence: sunhua@imm.ac.cn

Abstract: Background: Fumonisin B1 (FB1) is one of the most common mycotoxins contaminating feed and food. Although regulatory limits about fumonisins have been established in some countries, it is still very important to conduct research on lower doses of FB1 to determine the tolerance limits. The aim of this study was to investigate the effects of different concentrations of FB1, provide further evidence about the toxic doses- and exposure time-associated influence of FB1 on mice, especially low levels of FB1 for long-term exposure. Methods: Female BALB/c mice were treated intragastrically (i.g.) with fumonisin B1 (FB1) solutions (0 mg/kg body weight (BW), 0.018 mg/kg BW, 0.054 mg/kg BW, 0.162 mg/kg BW, 0.486 mg/kg BW, 1.458 mg/kg BW and 4.374 mg/kg BW) once a day for 8 weeks to obtain dose- and time-dependent effects on body and organ weights, hematology, blood chemical parameters and liver and kidney histopathology. Results: After the long-term administration of FB1, the body weights of the mice tended to decrease. Over time, FB1 first increased the relative spleen weight, then increased the relative kidney weight, and finally increased the relative liver weight. The mean corpuscular volume (MCV), mean corpuscular hemoglobin (MCH), hemoglobin (HGB), white blood cells (WBC), platelets (PLT), and mean platelet volume (MPV) were significantly elevated after treatment with FB1 for 8 weeks. Moreover, exposure time-dependent responses were found for aspartate aminotransferase (AST), alanine aminotransferase (ALT) and alkaline phosphatase (ALP) level, which were coupled with hepatic histopathological findings, necroinflammation and vacuolar degeneration and detrital necrosis. Linear dose response was also found for liver histopathology, in which, even the minimum dose of FB1 exposure also caused changes. Renal alterations were moderate compared to hepatic alterations. Conclusion: In conclusion, we demonstrated the systemic toxic effects of different doses of FB1 in female BALB/c mice at different times. Our data indicated that the effects observed in this study at the lowest dose tested are discussed in relation to the currently established provisional maximum tolerable daily intake (PMTDI) for fumonisins. This study suggested that recommendations for the concentration of FB1 in animals and humans are not sufficiently protective and that regulatory doses should be modified to better protect animal and human health. The toxicity of FB1 needs more attention.

Keywords: fumonisin B1; BALB/c mice; hepatotoxicity; nephrotoxicity; haematological toxicity; regulatory limit

Key Contribution: This study suggests that recommendations for the concentration of FB1 in animals and humans are not sufficiently protective and that regulatory doses should be modified to better protect animal and human health.

Citation: Chen, Z.; Zhang, F.; Jiang, L.; Chen, Z.; Sun, H. Toxic Effects of Mycotoxin Fumonisin B1 at Six Different Doses on Female BALB/c Mice. *Toxins* **2022**, *14*, 21. https://doi.org/10.3390/toxins14010021

Received: 3 December 2021
Accepted: 24 December 2021
Published: 29 December 2021

Publisher's Note: MDPI stays neutral with regard to jurisdictional claims in published maps and institutional affiliations.

1. Introduction

Mycotoxins, the secondary metabolites mainly produced by *Aspergillus, Penicillium,* and *Fusarium,* are highly poisonous substances in animals and humans. They are capable of causing mycotoxicosis [1], involving acute toxic, carcinogenic, mutagenic, teratogenic, immunotoxic, and estrogenic effects [2]. Since the discovery of the aflatoxins in the 1960s, an increasing number of mycotoxins have been characterized, including deoxynivalenol, T2 toxin, fumonisins, ochratoxin, and zearalenone [3].

Fumonisins (FBs) are a group of hydrophilic mycotoxins produced by *Fusarium verticillioides* and its related species that commonly contaminate corn, sorghum, related grains and even the traditional Chinese medicines (TCMs) throughout the world [4]. Until now, the fumonisins characterized since 1988 can be divided into four major groups as fumonisin A, B, C and P series [5]. Among these four groups, the most abundant and toxic fumonisin analog is fumonisin B1 (FB1), which contributes to approximately 70% of FBs and is one of the most common mycotoxins contaminating feed and food [6]. FB1 has been classified by the International Agency for Research on Cancer (IARC) as a Group 2B possibly carcinogenic to humans [2]. Regulatory limits on fumonisins have been established in some countries. In the European Union, the maximum level of total FBs (FB1 + FB2) range from 200 μg/kg in processed maize-based foods to 2000 μg/kg in unprocessed corn products [7]. The Food and Agriculture Organization/World Health Organization (FAO/WHO) specified a tolerable daily intake (TDI) of 2 μg/kg BW/day for fumonisins (FB1, FB2 and FB3, alone or by combination) [8].

More evidence indicates that FB1 is neurotoxic [9], nephrotoxic, hepatotoxic [6], hepatocarcinogenic [10] and immunotoxic [11]. As a potential hazardous contaminant, FB1 has been shown to cause the production of equine leukoencephalomalacia (ELEM) and porcine pulmonary edema (PPE) [4,12]. The association of the intake of fumonisins with human neural tube defects (NTDs) in the fetus has also been shown in regions where maize is consumed as a major food source [13]. It is also regarded as a high incidence of human esophageal cancer [12]. The contamination of food [14,15], feed [16,17] and traditional Chinese medicines (TCMs) [18] with fumonisins has been an increasingly serious concern in our society.

The aim of this study was to investigate the effects of different concentrations of FB1, including a concentration corresponding to the PMTDI (provisional maximum tolerable daily intake) of 2 μg/kg BW for FB1, FB2 and FB3, alone or in combination by the Joint FAO/WHO Expert Committee on Food Additives (JECFA) [8] on food. These six different concentrations of FB1 solutions (0.018 mg/kg BW, 0.054 mg/kg BW, 0.162 mg/kg BW, 0.486 mg/kg BW, 1.458 mg/kg BW and 4.374 mg/kg BW) were continuously administered to female BALB/c mice, which are more sensitive to the toxicity of FB1 [19,20].

2. Results

2.1. Effects of FB1 on the General Health and Body and Organ Weights of Mice

All mice were weighed once two weeks before being sacrificed. All of the body weights (BWs) of the mice are shown in Figure 1A. When mice were fed with FB1 for 2 weeks, body weights of FB1-5 (1.458 mg/kg BW) and FB1-6 (4.374 mg/kg BW) were higher than the control group, whereas after 4 weeks exposure, FB1-3 and FB1-4 body weight resulted notably lower when compared to the control. After 6 and 8 weeks of FB1 treatment, the body weights of mice in the FB1-6 (4.374 mg/kg BW) group were significantly lower than those in the control group.

As regards the organ weights, only spleen of FB1-5 (1.458 mg/kg BW) and FB1-6 (4.374 mg/kg BW) showed a significant increase in its weight after 2 weeks' exposure (Figure 1B). After 4 weeks, in addition to the relative spleen weights, the relative kidney weights of mice also presented an obvious increase in the high dose of FB1 (FB1-5 (1.458 mg/kg BW) and FB1-6 (4.374 mg/kg BW)). For mice treated with FB1 for 6 weeks, only the relative kidney weights in the FB1-6 group showed a significant increase compared with the control group. For mice treated with FB1 for 8 weeks, the relative kidney weights in the FB1-6 (4.374 mg/kg BW) group still showed a significant increase compared with the control group (Figure 1C). In addition, there was an obvious change in relative liver weight. The mice in all FB1-treated groups showed significantly higher relative liver weights than the control group (Figure 1D). From beginning to end, the relative heart weights of mice in each FB1-treated group showed no obvious changes compared to the control group (Figure 1E).

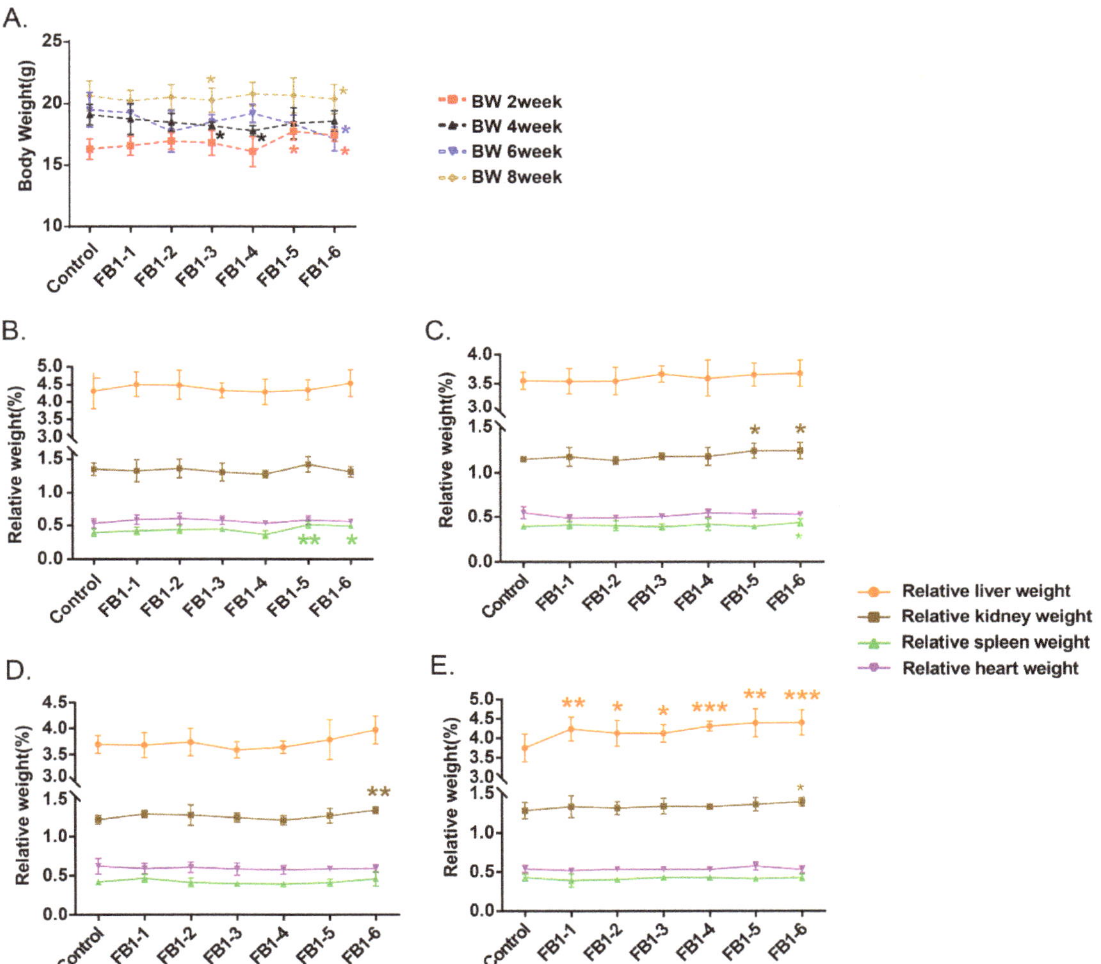

Figure 1. Effects of FB1 on body and organ weights of mice. (**A**) Mean body weight of mice exposed to different levels of FB1 for 2, 4, 6, and 8 consecutive weeks. (**B–E**) Relative organ weights of mice exposed to different levels of FB1 for 2, 4, 6 and 8 weeks. Values are shown as the mean ± SEM; 2 weeks ($n = 5$), 4 weeks ($n = 6$–7), 6 weeks ($n = 5$), and 8 weeks ($n = 9$). FB1-1 indicates 0.018 mg/kg BW, FB1-2 indicates 0.054 mg/kg BW, FB1-3 indicates 0.162 mg/kg BW, FB1-4 indicates 0.486 mg/kg BW, FB1-5 indicates 1.458 mg/kg BW and FB1-6 indicates groups 4.374 mg/kg BW; * indicates $p < 0.05$, ** indicates $p < 0.01$, *** indicates $p < 0.001$, vs. the control group.

2.2. Effects of FB1 on Hematology of Mice

After 8 weeks of administration of FB1, the changes of red blood cells (RBC) and related parameters were shown in Figure 2A–G. Compared with the control group, the levels of mean corpuscular volume (MCV) and hemoglobin (HGB) in all FB1-treated groups were significantly increased (Figure 2C,G). Mean corpuscular hemoglobin (MCH) also presented a significant increase in the FB1-2 (0.054 mg/kg BW), FB1-3 (0.162 mg/kg BW), FB1-4 (0.486 mg/kg BW), FB1-5 (1.458 mg/kg BW) and FB1-6 groups (4.374 mg/kg BW) (Figure 2D). For red cell volume distribution width (RDW), its level was decreased in the FB1-2 (0.054 mg/kg BW), FB1-3 (0.162 mg/kg BW), FB1-4 (0.486 mg/kg BW) and FB1-5 (1.458 mg/kg BW) groups (Figure 2F). However, there was no significant difference in the levels of RBC, red blood cell specific volume (HCT) and mean corpuscular hemoglobin concentration (MCHC) (Figure 2A,B,E).

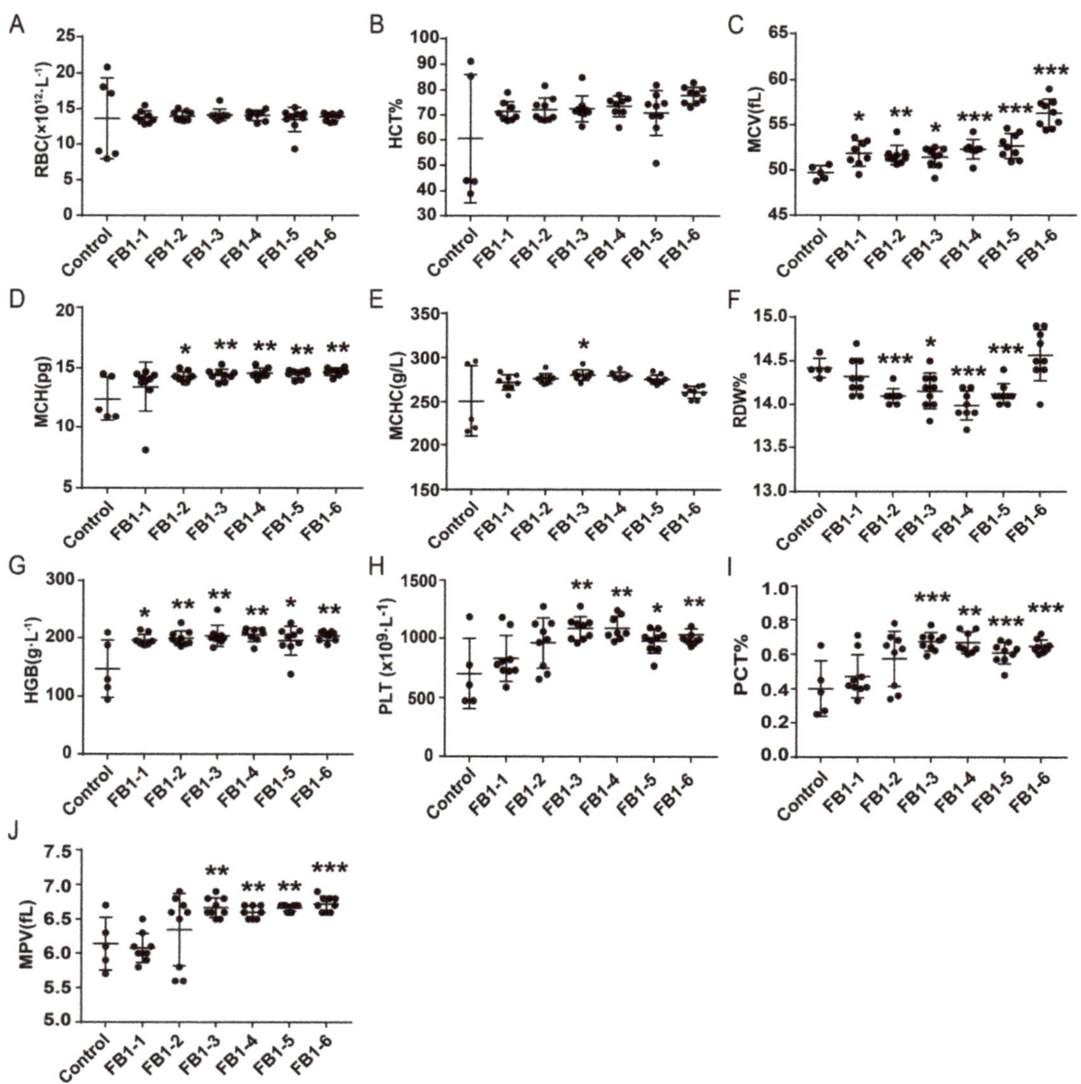

Figure 2. *Cont.*

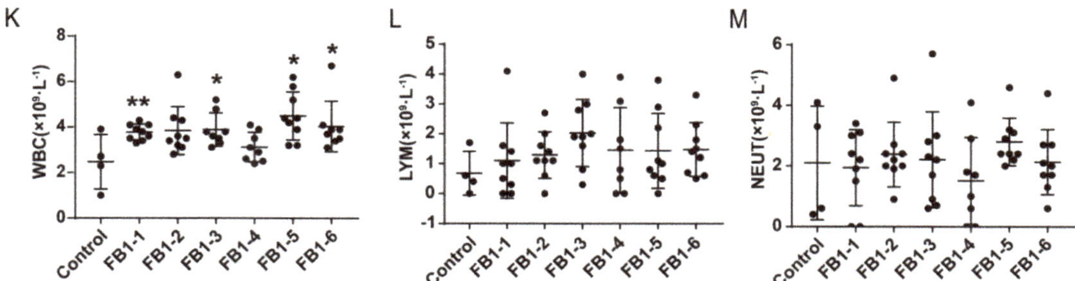

Figure 2. Effects of FB1 on hematology and related parameters in mice after 8 weeks of administration of FB1. (**A–G**): The changes of red blood cells (RBC) and related parameters. (**H–J**): The changes of PLT and related parameters. K to M: The changes of white blood cells (WBC) and related parameters. Values are shown as the mean $\pm$ SEM ($n = 9$); 2 weeks ($n = 5$), 4 weeks ($n = 6$–7), 6 weeks ($n = 5$), and 8 weeks ($n = 9$). FB1-1 indicates 0.018 mg/kg BW, FB1-2 indicates 0.054 mg/kg BW, FB1-3 indicates 0.162 mg/kg BW, FB1-4 indicates 0.486 mg/kg BW, FB1-5 indicates 1.458 mg/kg BW and FB1-6 indicates groups 4.374 mg/kg BW; * indicates $p < 0.05$, ** indicates $p < 0.01$, *** indicates $p < 0.001$, vs. the control group.

There were also some significant changes in the levels of PLT and related parameters after treatment with FB1 for 8 weeks. In the FB1-3 (0.162 mg/kg BW), FB1-4 (0.486 mg/kg BW), FB1-5 (1.458 mg/kg BW) and FB1-6 groups (4.374 mg/kg BW), platelet (PLT), plateletocrit (PCT) and mean platelet volume (MPV) were higher than those of the control group (Figure 2H–J).

The number of white blood cells (WBC) showed a dramatic increase in the FB1-1 (0.018 mg/kg BW), FB1-3 (0.162 mg/kg BW), FB1-5 (1.458 mg/kg BW) and FB1-6 (4.374 mg/kg BW) groups (Figure 2K), but the levels of lymphocytes (LYM) and neutrophils (NEUT) had no changes (Figure 2L,M).

2.3. Effects of FB1 on Blood Chemistry Parameters of Mice

As shown in Figure 3, the serum AST, ALT and ALP activities showed dose-associated and exposure time-associated increases. All three enzymes reached peak values in the FB1-6 (4.374 mg/kg BW) group, especially when FB1 was given for 8 weeks. Even the value of AST in the FB1-2 (0.054 mg/kg BW) group after 8 weeks of FB1 treatment also presented a significant increase. The ALP level increased significantly at the FB1-6 dose after FB1 was administered to mice for only 2 weeks.

For blood urea nitrogen (BUN) and creatinine (CRE), during the first 4 weeks of exposure to FB1, the levels of BUN in the FB1-5 (1.458 mg/kg BW) and FB1-6 (4.374 mg/kg BW) groups were decreased significantly compared with those in the control group, whereas at the end of the experiment, BUN in the FB1-5 (1.458 mg/kg BW) group was obviously increased. The fluctuation of the CRE level during the whole experiment was obvious. CRE in mice showed a significantly decreased tendency except for the FB1-1 group after 2 weeks of exposure to FB1.

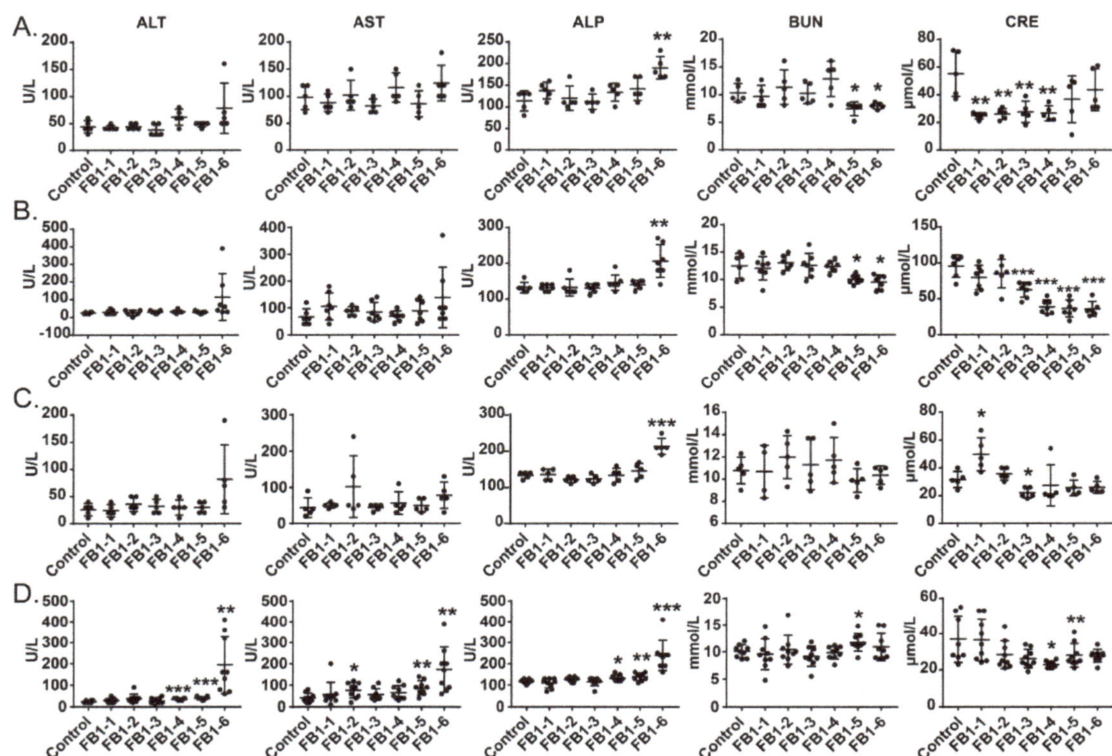

Figure 3. (**A–D**) Effects of FB1 on the blood chemistry parameters of mice exposed to different levels of FB1 for 2, 4, 6 and 8 weeks. Values are shown as the mean $\pm$ SEM; 2 weeks (n = 5), 4 weeks (n = 6–7), 6 weeks (n = 5), and 8 weeks (n = 9). FB1-1 indicates 0.018 mg/kg BW, FB1-2 indicates 0.054 mg/kg BW, FB1-3 indicates 0.162 mg/kg BW, FB1-4 indicates 0.486 mg/kg BW, FB1-5 indicates 1.458 mg/kg BW and FB1-6 indicates groups 4.374 mg/kg BW; * indicates $p < 0.05$, ** indicates $p < 0.01$, *** indicates $p < 0.001$, vs. the control group.

2.4. Effects of FB1 on Histopathologic Changes

2.4.1. Liver

Histological changes were found in the livers of mice in all FB1-treated groups. The main histological lesions observed in the liver were necroinflammation, including periportal necrosis and inflammation, intralobular necrosis and inflammation as well as portal inflammation. As shown in Figure 4, in the group of mice exposed to FB1-1 (0.018 mg/kg BW) and (0.054 mg/kg BW) for 2 weeks, infiltration of scattered inflammatory cells around the manifold area was already found. In the FB1-3 (0.162 mg/kg BW) group, a small number of hepatocytes started to exhibit necrosis. The effects were more severe in the FB1-5 (1.458 mg/kg BW) and FB1-6 (4.374 mg/kg BW) groups, as there were hepatocytes with vacuolar degeneration and detrital necrosis. From the results of lesional scores at 4, 6, and 8 weeks (Figures 4–7), we also found that the lesional scores of all FB1-treated groups were significantly higher than those of the control group, even in the FB1-1 (0.018 mg/kg BW) group with the minimum dose of FB1. The lesional score increased in parallel with the increasing exposure time of FB1, and the highest score was found in the FB1-6 (4.374 mg/kg BW) group after 8 weeks of exposure to FB1.

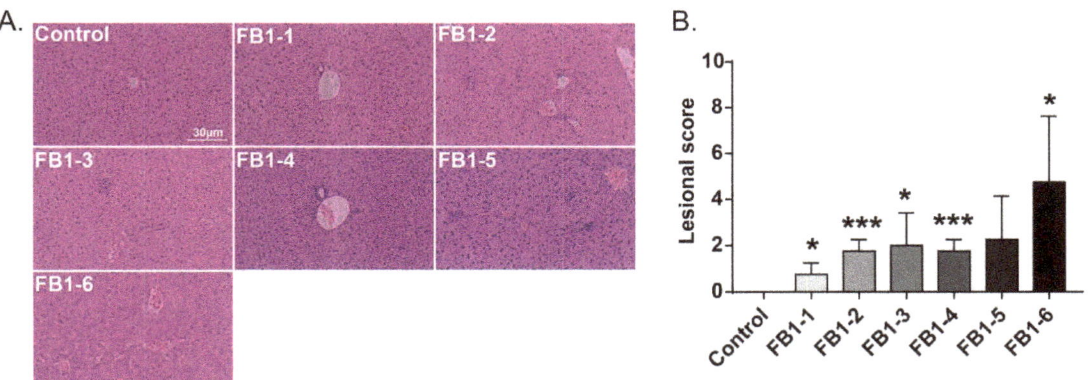

Figure 4. Histological changes in the liver in mice exposed to different doses of FB1 for 2 weeks. (**A**) Typical histological picture in each group (HE. 200×). (**B**) Lesional score after histological examination based on the occurrence and severity of lesions. Values are shown as the mean ± SEM (n = 5). FB1-1 indicates 0.018 mg/kg BW, FB1-2 indicates 0.054 mg/kg BW, FB1-3 indicates 0.162 mg/kg BW, FB1-4 indicates 0.486 mg/kg BW, FB1-5 indicates 1.458 mg/kg BW and FB1-6 indicates groups 4.374 mg/kg BW; * means $p < 0.05$, *** means $p < 0.001$, vs. control group.

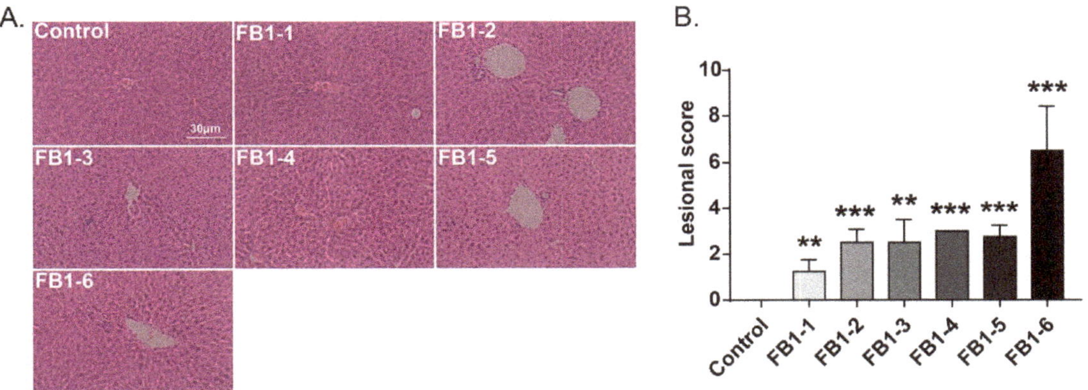

Figure 5. Histological changes in the liver in mice exposed to different doses of FB1 for 4 weeks. (**A**) Typical histological picture in each group (HE. 200×). (**B**) Lesional score after histological examination based on the occurrence and severity of lesions. Values are shown as the mean ± SEM (n = 6–7). FB1-1 indicates 0.018 mg/kg BW, FB1-2 indicates 0.054 mg/kg BW, FB1-3 indicates 0.162 mg/kg BW, FB1-4 indicates 0.486 mg/kg BW, FB1-5 indicates 1.458 mg/kg BW and FB1-6 indicates groups 4.374 mg/kg BW; * indicates $p < 0.05$, ** indicates $p < 0.01$, *** indicates $p < 0.001$, vs. the control group.

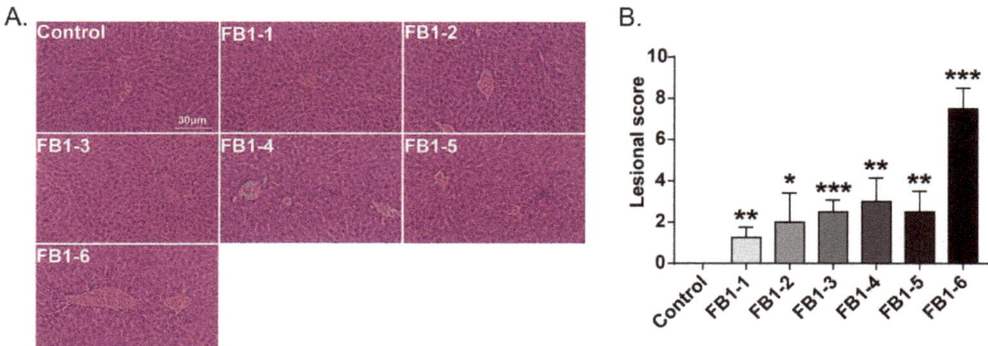

Figure 6. Histological changes in the liver in mice exposed to different doses of FB1 for 6 weeks. (**A**) Typical histological picture in each group (HE. 200×). (**B**) Lesional score after histological examination based on the occurrence and severity of lesions. Values are shown as the mean ± SEM ($n = 5$). FB1-1 indicates 0.018 mg/kg BW, FB1-2 indicates 0.054 mg/kg BW, FB1-3 indicates 0.162 mg/kg BW, FB1-4 indicates 0.486 mg/kg BW, FB1-5 indicates 1.458 mg/kg BW and FB1-6 indicates groups 4.374 mg/kg BW; * indicates $p < 0.05$, ** indicates $p < 0.01$, *** indicates $p < 0.001$, vs. control group.

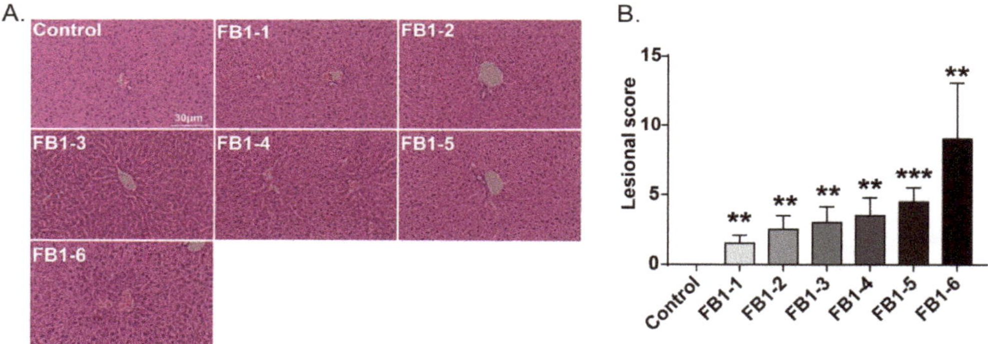

Figure 7. Histological changes in the liver in mice exposed to different doses of FB1 for 8 weeks. (**A**) Typical histological picture in each group (HE. 200×). (**B**) Lesional score after histological examination based on the occurrence and severity of lesions. Values are shown as the mean ± SEM ($n = 9$). FB1-1 indicates 0.018 mg/kg BW, FB1-2 indicates 0.054 mg/kg BW, FB1-3 indicates 0.162 mg/kg BW, FB1-4 indicates 0.486 mg/kg BW, FB1-5 indicates 1.458 mg/kg BW and FB1-6 indicates groups 4.374 mg/kg BW; * indicates $p < 0.05$, ** indicates $p < 0.01$, *** indicates $p < 0.001$, vs. control group.

2.4.2. Kidneys

In mice treated with different FB1 concentrations for different time periods, the main histological lesions observed in the kidneys were glomerular injury, renal tubular injury, and renal interstitial inflammation. Typical histological pictures are shown in Figure 8, and the corresponding lesional scores are shown in Table 1. Overall, vacuolar degeneration of renal tubular epithelial cells was ubiquitous in all FB1-treated groups and worsened over time. There was also tubular atrophy in some kidneys. For glomerular injury, slight mesangial cell hyperplasia first appeared in mice in the FB1-4 (0.486 mg/kg BW) group after exposure to FB1 for 4 weeks (Figure 8B), and the worst case occurred in FB1-6 (4.374 mg/kg BW) after the 8 week experiment (Figure 8D and Table 1).

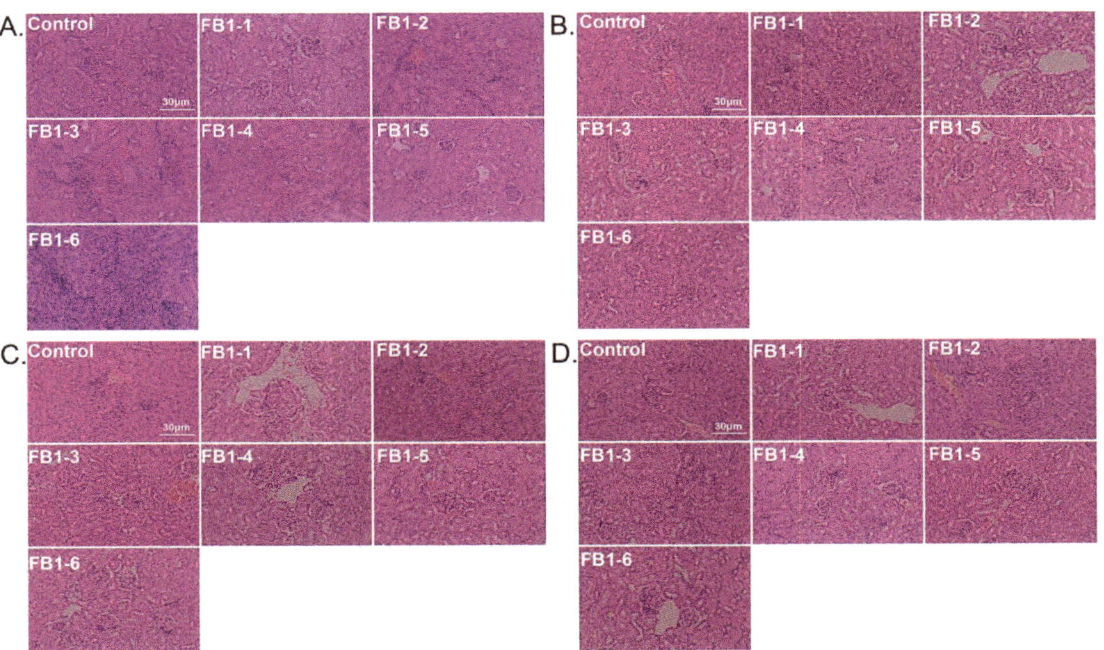

Figure 8. Histological changes in the kidneys in mice (HE. 200×). (**A–D**). Typical histological picture of mice suffering from different doses of FB1 for 2, 4, 6 and 8 weeks; 2 weeks ($n = 5$), 4 weeks ($n = 6$–7), 6 weeks ($n = 5$), and 8 weeks ($n = 9$). FB1-1 indicates 0.018 mg/kg BW, FB1-2 indicates 0.054 mg/kg BW, FB1-3 indicates 0.162 mg/kg BW, FB1-4 indicates 0.486 mg/kg BW, FB1-5 indicates 1.458 mg/kg BW and FB1-6 indicates groups 4.374 mg/kg BW.

Table 1. Lesional score of the kidneys in mice exposed to different levels of FB1 for 2, 4, 6 and 8 weeks.

2 Weeks	Control	FB1-1	FB1-2	FB1-3	FB1-4	FB1-5	FB1-6
Glomerulus	0.00 ± 0.00	0.00 ± 0.00	0.33 ± 0.58	0.00 ± 0.00	0.00 ± 0.00	0.00 ± 0.00	0.67 ± 0.58
Renal tubule	0.00 ± 0.00	0.33 ± 0.58	0.67 ± 0.58	1.00 ± 0.00	1.00 ± 1.00	0.67 ± 0.58	1.00 ± 0.00
Intestitial inflammation	0.00 ± 0.00	0.00 ± 0.00	1.33 ± 0.58 *	1.33 ± 0.58 *	0.67 ± 0.58	1.00 ± 0.00	2.00 ± 0.00

4 Weeks	Control	FB1-1	FB1-2	FB1-3	FB1-4	FB1-5	FB1-6
Glomerulus	0.00 ± 0.00	0.00 ± 0.00	0.00 ± 0.00	0.00 ± 0.00	0.33 ± 0.58	0.00 ± 0.00	0.00 ± 0.00
Renal tubule	0.00 ± 0.00	0.67 ± 0.58	1.00 ± 0.00	1.00 ± 0.00	0.67 ± 0.58	1.33 ± 0.58 *	0.67 ± 0.58
Intestitial inflammation	0.00 ± 0.00	0.00 ± 0.00	1.00 ± 0.00	0.67 ± 0.58	0.67 ± 0.58	0.33 ± 0.58	1.00 ± 0.00

6 Weeks	Control	FB1-1	FB1-2	FB1-3	FB1-4	FB1-5	FB1-6
Glomerulus	0.00 ± 0.00	0.00 ± 0.00	0.00 ± 0.00	0.67 ± 0.58	0.67 ± 0.58	0.67 ± 0.58	0.67 ± 0.58
Renal tubule	0.00 ± 0.00	0.67 ± 0.58	0.33 ± 0.58	1.00 ± 0.00	1.33 ± 0.58 *	0.67 ± 0.58	1.33 ± 0.58 *
Intestitial inflammation	0.00 ± 0.00	0.33 ± 0.58	1.00 ± 0.00	0.67 ± 0.58	0.00 ± 0.00	0.33 ± 0.58	0.67 ± 0.58

Table 1. *Cont.*

8 Weeks	Control	FB1-1	FB1-2	FB1-3	FB1-4	FB1-5	FB1-6
Glomerulus	0.00 ± 0.00	0.00 ± 0.00	0.67 ± 0.58	0.67 ± 0.58	1.00 ± 0.00	1.00 ± 0.00	1.00 ± 0.00
Renal tubule	0.00 ± 0.00	1.00 ± 0.00	0.67 ± 0.58	1.33 ± 0.58 *	1.00 ± 0.00	1.00 ± 0.00	1.67 ± 0.58 *
Intestitial inflammation	0.00 ± 0.00	0.00 ± 0.00	0.33 ± 0.58	0.67 ± 0.58	0.00 ± 0.00	0.33 ± 0.58	1.00 ± 0.00

Values are shown as the mean $\pm$ SEM; 2 weeks (n = 5), 4 weeks (n = 6–7), 6 weeks (n = 5), and 8 weeks (n = 9). FB1-1 indicates 0.018 mg/kg BW, FB1-2 indicates 0.054 mg/kg BW, FB1-3 indicates 0.162 mg/kg BW, FB1-4 indicates 0.486 mg/kg BW, FB1-5 indicates 1.458 mg/kg BW and FB1-6 indicates groups 4.374 mg/kg BW; Values are shown as the mean $\pm$ SEM; 2 weeks (n = 5), 4 weeks (n = 6–7), 6 weeks (n = 5), and 8 weeks (n = 9); * indicates $p < 0.05$ vs. control group.

3. Discussion

The liver and kidneys are the main target organs for the toxic effects of FB1 [21,22]. In this study, female BALB/c mice were exposed to different concentrations of FB1 by intragastric administration based on the fact that FB1 was more frequently ingested by the mouth. The liver and kidney toxicity of FB1 in mice was examined from two dimensions: dose and exposure time. Through the transformation of body surface area based on animal basal metabolic rate, the dose of 20 g mice is about 12.3 times that of 60 kg adults [23]. Therefore, the dose of FB1-1 (0.018 mg/kg BW) corresponds to the PMTDI (2 μg/kg BW) of FB1, FB2 and FB3, which is applied by the Joint FAO/WHO Expert Committee on Food Additives (JECFA). The results of this study suggested that the limit standard formulated by the FAO/WHO and the U.S. Food and Drug Administration (USFDA) is insufficient to protect humans and animals.

3.1. Body and Organ Weights

Since mice were exposed to FB1 for 4 weeks, there was a decreasing tendency in the body weights of mice, which reflected the cumulative alterations of FB1 treatment. Interestingly, exposure of FB1 for two weeks, high doses of FB1 (1.458 mg/kg BW and 4.374 mg/kg BW) actually increased the animal's body weight. Previous studies have shown that low-dose aflatoxin is a hormesis for chickens [24]. Whether the short-term exposure of FB1 has the effect of increasing body weight may require further research. Changes in relative organ weights can better reflect the animal's state and be reliable indicators of some physiological changes [25,26]. The results of this study showed that FB1 exposure had no significant effect on the relative heart weights of mice. However, the relative spleen weights increased first (2 weeks), and then the relative kidney weights increased (4 weeks), and finally the relative liver weights increased (8 weeks). The long-term effect of the minimum dose of FB1-1 (0.018 mg/kg BW) could also cause liver damage. The results of relative organ weights also suggested that FB1 might damage the immune system first and then damage the kidneys, finally leading to liver damage as the exposure time increased. This finding requires further research for confirmation.

3.2. Blood Chemistry

ALT and AST in serum are important indicators for reflecting liver function. ALT is an active enzyme mainly present in the liver, and AST has the highest content in the myocardium, followed by the liver. Both ALT and AST are released from damaged hepatocytes into the blood after hepatocellular injury or death [27]. ALP is an important indicator of cholestasis and hepatobiliary diseases in humans. Serum ALP levels can be elevated by cholestatic or infiltrative diseases of the liver and by diseases causing obstruction to the biliary system [27]. The results of this study showed that when FB1 was administered to mice for 8 consecutive weeks at the FB1-4 (0.486 mg/kg BW) dose, which was lower than the 4 mg/kg BW limit in corn flour and its products regulated by the USFDA, the ALT and ALP of mice was also significantly higher than that in the control group. The above results suggested that the dose of FB1 that is lower than the 4 mg/kg limit in corn flour and its products regulated by the USFDA has induced significant alterations in biochemical indicators.

3.3. Hematology

Routine blood tests are the most accessible and fundamental examination and have long been proposed as an essential assistant tool for disease diagnosis [28]. Changes in the blood system can predict the occurrence of some diseases. From our results, there were some significant hematological changes in female BALB/c mice after exposure to FB1. In particular, the MCV of mice in the low-dose FB1-1 (0.018 mg/kg BW) group changed significantly. The WBCs were estimated as an indicator of the immune response. Thus, Khawla Ezdini suggested that an inflammatory reaction to fight against mycotoxins impaired immunity [29]. In addition, MPV is a platelet volume index directly reflecting the platelet function state [30] and its elevation usually leads to thrombocytosis [29]. In our results, a significant increase in the number of WBCs existed in the low-dose FB1-1 (0.018 mg/kg BW) group and a significant increase in the number of PLTs and related parameters PCT and MPV in the low-dose FB1-3 (0.162 mg/kg BW) group. The results of hematology also suggested that the limit standard formulated by the FAO/WHO and USFDA is insufficient to protect humans and animals.

3.4. Histopathological Analysis

Exposure to FB1 resulted in injuries to the liver and kidneys, which could be directly reflected by histopathological analysis. After exposure to FB1 for 2 weeks, liver histological alterations were already observed at a dose of FB1-1 (0.018 mg/kg BW), which is lower than the limit standard formulated by the FAO/WHO. In terms of population, food consumption risk assessments should be performed in order to question the limit set by the FAO/WHO. In addition, the severity of lesions in the liver increased progressively, and dose-dependent severity was observed. The degree of lesions in the kidneys was not as severe as liver histological alterations, which was consistent with the finding that female mice were more sensitive to the hepatotoxic effects of FB1 than their male counterparts [31].

4. Conclusions

FB1 can cause significant hepatotoxicity, nephrotoxicity and hematological toxicity. Renal toxicity precedes hepatotoxicity, and the toxicity exhibits a certain dose dependence and exposure time dependence, especially histological alterations in the liver. The limit of the maximum tolerable daily intake of fumonisin in foods set by the FAO/WHO of 2 µg/kg body weight/day does not seem to have sufficient protection. FB1-1 (0.018 mg/kg BW), which is lower than this limit based on the dosage transformation of body surface area, can cause obvious deleterious influences on the liver and kidneys in female BALB/c mice, such as hepatocyte degeneration, necrosis and inflammation in the manifold and renal tubular damage in the kidneys. The FB1-4 dose (0.486 mg/kg BW) is slightly lower than the limit made by the USFDA for fumonisins in cornmeal and its products, and the FB1-1 (0.018 mg/kg BW), FB1-2 (0.054 mg/kg BW), and FB1-3 (0.162 mg/kg BW) doses are all significantly lower than this limit. These four doses of FB1 showed more significant toxic effects on mice. In short, the current regulatory limits for fumonisins are not sufficiently protective. Therefore, there would be more interest and importance in conducting further studies to determine the levels with the lack of observed adverse effects for fumonisins.

5. Materials and Methods

5.1. FB1 Solution Preparation

The FB1 solution was prepared by dissolving FB1 powder (Pribolab, Qingdao, China) in distilled water. First, we obtained a 0.4374 mg/mL FB1 solution and then diluted it with distilled water to reach concentrations of 0.1458 mg/mL, 0.0486 mg/mL, 0.0162 mg/mL, 0.0054 mg/mL and 0.0018 mg/mL.

5.2. Animal Trial

In this study, 181 female BALB/c mice (SPF grade, HFK Bioscience Co., Ltd. Beijing, China) with a body weight of 16–18 g and no specific pathogens (SPF) were used. The mice were kept in the Animal Experimental Center, Institute of Materia Medica, CAMS & PUMC with no restricted access to commercial feed and water. There were 4–5 mice kept in each cage. They were maintained in an environment at 20 ± 3 °C, 50.0 ± 10.0% humidity and a 12-h light/dark cycle. After acclimated for one week the mice were randomly divided into 7 groups according to body weight, which were labeled as the FB1-1 (0.018 mg/kg BW, 26 mice), FB1-2 (0.054 mg/kg BW, 26 mice), FB1-3 (0.162 mg/kg BW, 26 mice), FB1-4 (0.486 mg/kg BW, 26 mice), FB1-5 (1.458 mg/kg BW, 26 mice), FB1-6 (4.374 mg/kg BW, 26 mice), and the control group (25 mice). The doses of FB1-4 (0.486 mg/kg BW) and FB1-5 (1.458 mg/kg BW) crossed over the recommended levels of 4 mg/kg BW for total fumonisins (FB1 + FB2 + FB3) in whole or partially degermed dry milled corn products (e.g., flaking grits, corn grits, corn meal, corn flour with fat content of >2.25%, dry weight basis) by the USFDA. The control group was gavage with distilled water, and the other groups were gavage with corresponding FB1 solution. The feeding volume was 10 mL/kg per mouse each time and once a day. These similar treatments lasted for 2 ($n = 5$), 4 ($n = 6$–7), 6 ($n = 5$), and 8 ($n = 9$) weeks. On the last day of the 2, 4, 6, and 8 week periods, the mice were anesthetized by pentobarbital (80 mg/kg BW, given intraperitoneally (i.p.)). After collecting blood from the retro-orbital plexus, the animals were euthanasia by cervical dislocation and immediately dissected. The study was conducted according to the guidelines of the Declaration of Helsinki, and approved by The Animal Care & Welfare Committee, Institute of Materia Medica, CAMS & PUMC (protocol code 00003407. Approval Date: 30 October 2018).

5.3. Hematological Analysis

Exactly 20 μL of blood for hematological analysis were dispensed into test tubes containing anticoagulant. This was used for whole-blood analysis using an automatic blood analyzer following the instruction manual. The red blood cells (RBC), haemoglobin (HGB) red blood cell specific volume (HCT), mean corpuscular volume (MCV), mean corpuscular haemoglobin (MCH), mean corpuscular haemoglobin concentration (MCHC), RBC distribution width (RDW), white blood cells (WBC), lymphocyte (LYM), neutrophil (NEUT), platelets (PLT), plateletocrit (PCT), mean platelet volume (MPV) and platelet distribution width (PDW) were determined.

5.4. Blood Chemistry Parameters Analysis

Whole blood from each mouse was centrifuged at 4000 rpm for 15 min to prepare serum. The serum aspartate aminotransferase (AST) activity, alanine aminotransferase (ALT) activity, alkaline phosphatase (ALP) activity, creatinine (CRE), and urea nitrogen (BUN) were determined by a TBA-40FR Chemistry Analyzer (TOSHIBA, Tokyo, Japan) with commercial diagnostic kits (Biosino Bio-Technology and Science Inc., Beijing, China).

5.5. Histopathological Analysis

After gross examination of the organs of mice at necropsy, the liver and kidneys were fixed with 4% paraformaldehyde buffer, embedded in paraffin and blocked. The tissue blocks were sectioned to 4 μm and stained with hematoxylin-eosin for light microscopic analysis.

For the liver, histopathological analysis was performed according to the Knodell grading system, and the histological activity index (HAI) (range, 0–18) was used to grade the histological changes in tissues [32]. The overall Knodell score is the sum of scores for periportal ± bridging necrosis (0–10), intralobular degeneration and focal necrosis (0–4), portal inflammation (0–4), and fibrosis (0–4). The lesional score in our results is the sum of the first 3 components, which is used to reflect the necroinflammatory activity index. For the kidneys, the main pathological alterations were described and scored according to their extent and severity as follows: 0 indicated that a lesion was not present and 1–3 indicated that lesions were slight, moderate and severe.

5.6. Statistical Analysis

Statistical significance was determined following the test using either one-tailed *t*-test or one-way analysis of variance (ANOVA) with Tukey's multiple comparisons to compare the means of multiple groups. Data are shown as the mean $\pm$ SEM and were considered statistically significant at $p < 0.05$. GraphPad Prism 7 (GraphPad Software Inc., LaJolla, CA, USA) was used for analysis and graphic building.

Author Contributions: Funding acquisition, H.S.; Investigation, Z.C. (Zhiwei Chen), F.Z., L.J. and Z.C. (Zihan Chen).; Methodology, Z.C. (Zhiwei Chen), F.Z. and H.S.; Project administration, H.S.; Resources, H.S.; Validation, H.S.; Visualization, Z.C. (Zhiwei Chen) and F.Z.; Writing—original draft, Z.C. (Zhiwei Chen) and F.Z.; Writing—review & editing, Z.C. (Zhiwei Chen). All authors have read and agreed to the published version of the manuscript.

Funding: This work was funded by the National Key Research and Development Program of China [No. 2019YFC1708901], the CAMS Innovation Fund for Medical Sciences (grant number 2017-I2M-1-013), the National Science & Technology Major Project "Key New Drug Creation and Manufacturing Program" (grant number 2019ZX09201001), and the Drug Innovation Major Project (grant number 2018ZX09711001-003-011).

Institutional Review Board Statement: The study was conducted according to the guidelines of the Declaration of Helsinki, and approved by The Animal Care & walfare Committe Instltute of Materica Medica, CAMS & PUMC. (protocol code 00003407. Approval Date: 30 October 2018).

Informed Consent Statement: Not applicable.

Data Availability Statement: Data available on request due to restrictions e.g., privacy or ethical.

Conflicts of Interest: The authors declare no conflict of interest.

References

1. Liew, W.P.; Mohd-Redzwan, S. Mycotoxin: Its Impact on Gut Health and Microbiota. *Front. Cell Infect. Microbiol.* **2018**, *8*, 60. [CrossRef] [PubMed]
2. Cunha, S.C.; Sá, S.V.M.; Fernandes, J.O. Multiple mycotoxin analysis in nut products: Occurrence and risk characterization. *Food Chem. Toxicol.* **2018**, *114*, 260–269. [CrossRef] [PubMed]
3. Janik, E.; Niemcewicz, M.; Ceremuga, M.; Stela, M.; Saluk-Bijak, J.; Siadkowski, A.; Bijak, M. Molecular Aspects of Mycotoxins-A Serious Problem for Human Health. *Int. J. Mol. Sci.* **2020**, *21*, 8187. [CrossRef] [PubMed]
4. Kamle, M.; Mahato, D.K.; Devi, S.; Lee, K.E.; Kang, S.G.; Kumar, P. Fumonisins: Impact on Agriculture, Food, and Human Health and their Management Strategies. *Toxins* **2019**, *11*, 328. [CrossRef] [PubMed]
5. Rheeder, J.P.; Marasas, W.F.; Vismer, H.F. Production of fumonisin analogs by Fusarium species. *Appl. Environ. Microbiol.* **2002**, *68*, 2101–2105. [CrossRef] [PubMed]
6. Singh, M.P.; Kang, S.C. Endoplasmic reticulum stress-mediated autophagy activation attenuates fumonisin B1 induced hepatotoxicity in vitro and in vivo. *Food Chem. Toxicol.* **2017**, *110*, 371–382. [CrossRef]
7. Knutsen, H.K.; Alexander, J.; Barregård, L.; Bignami, M.; Brüschweiler, B.; Ceccatelli, S.; Cottrill, B.; Dinovi, M.; Edler, L.; Grasl-Kraupp, B.; et al. Risks for animal health related to the presence of fumonisins, their modified forms and hidden forms in feed. *EFSA J.* **2018**, *16*, e05242. [CrossRef]
8. Joint FAO/WHO Expert Committee on Food Additives Eighty-Third Meeting. Available online: http://www.fao.org/3/bq821e/bq821e.pdf (accessed on 20 August 2021).
9. Domijan, A.M. Fumonisin B(1): A neurotoxic mycotoxin. *Arch. Ind. Hyg. Toxicol.* **2012**, *63*, 531–544. [CrossRef]
10. Fukuda, H.; Shima, H.; Vesonder, R.F.; Tokuda, H.; Nishino, H.; Katoh, S.; Tamura, S.; Sugimura, T.; Nagao, M. Inhibition of protein serine/threonine phosphatases by fumonisin B1, a mycotoxin. *Biochem. Biophys. Res. Commun.* **1996**, *220*, 160–165. [CrossRef]
11. Dombrink-Kurtzman, M.A.; Gomez-Flores, R.; Weber, R.J. Activation of rat splenic macrophage and lymphocyte functions by fumonisin B1. *Immunopharmacology* **2000**, *49*, 401–409. [CrossRef]
12. Dutton, M.F. Fumonisins, mycotoxins of increasing importance: Their nature and their effects. *Pharmacol. Ther.* **1996**, *70*, 137–161. [CrossRef]
13. Japan, F.S.C.O. Fumonisins (Natural Toxins and Mycotoxins). *Food Saf.* **2018**, *6*, 160–161. [CrossRef]
14. Cendoya, E.; Nichea, M.J.; Monge, M.P.; Sulyok, M.; Chiacchiera, S.M.; Ramirez, M.L. Fumonisin occurrence in wheat-based products from Argentina. *Food Addit. Contam. Part B* **2019**, *12*, 31–37. [CrossRef] [PubMed]
15. Jiang, D.; Li, F.; Zheng, F.; Zhou, J.; Li, L.; Shen, F.; Chen, J.; Li, W. Occurrence and dietary exposure assessment of multiple mycotoxins in corn-based food products from Shandong, China. *Food Addit. Contam. Part B* **2019**, *12*, 10–17. [CrossRef]

16. Akinmusire, O.O.; El-Yuguda, A.D.; Musa, J.A.; Oyedele, O.A.; Sulyok, M.; Somorin, Y.M.; Ezekiel, C.N.; Krska, R. Mycotoxins in poultry feed and feed ingredients in Nigeria. *Mycotoxin Res.* **2019**, *35*, 149–155. [CrossRef]
17. Shao, M.; Li, L.; Gu, Z.; Yao, M.; Xu, D.; Fan, W.; Yan, L.; Song, S. Mycotoxins in commercial dry pet food in China. *Food Addit. Contam. Part B* **2018**, *11*, 237–245. [CrossRef]
18. Han, Z.; Ren, Y.; Liu, X.; Luan, L.; Wu, Y. A reliable isotope dilution method for simultaneous determination of fumonisins B1, B2 and B3 in traditional Chinese medicines by ultra-high-performance liquid chromatography-tandem mass spectrometry. *J. Sep. Sci.* **2010**, *33*, 2723–2733. [CrossRef]
19. He, Q.; Riley, R.T.; Sharma, R.P. Myriocin prevents fumonisin B1-induced sphingoid base accumulation in mice liver without ameliorating hepatotoxicity. *Food Chem. Toxicol.* **2005**, *43*, 969–979. [CrossRef]
20. Sozmen, M.; Devrim, A.K.; Tunca, R.; Bayezit, M.; Dag, S.; Essiz, D. Protective effects of silymarin on fumonisin B_1-induced hepatotoxicity in mice. *J. Vet. Sci.* **2014**, *15*, 51–60. [CrossRef]
21. Gruber-Dorninger, C.; Novak, B.; Nagl, V.; Berthiller, F. Emerging Mycotoxins: Beyond Traditionally Determined Food Contaminants. *J. Agric. Food Chem.* **2017**, *65*, 7052–7070. [CrossRef] [PubMed]
22. Stockmann-Juvala, H.; Savolainen, K. A review of the toxic effects and mechanisms of action of fumonisin B1. *Hum. Exp. Toxicol.* **2008**, *27*, 799–809. [CrossRef] [PubMed]
23. Nair, A.; Morsy, M.A.; Jacob, S. Dose translation between laboratory animals and human in preclinical and clinical phases of drug development. *Drug Dev. Res.* **2018**, *79*, 373–382. [CrossRef]
24. Diaz, G.J.; Calabrese, E.; Blain, R. Aflatoxicosis in chickens (Gallus gallus): An example of hormesis? *Poult. Sci.* **2008**, *87*, 727–732. [CrossRef] [PubMed]
25. Cintra, L.T.; Samuel, R.O.; Prieto, A.K.; Sumida, D.H.; Dezan-Júnior, E.; Gomes-Filho, J.E. Oral health, diabetes, and body weight. *Arch. Oral Biol.* **2017**, *73*, 94–99. [CrossRef] [PubMed]
26. Fan, L.; He, Z.Z.; Ao, X.; Sun, W.L.; Xiao, X.; Zeng, F.K.; Wang, Y.C.; He, J. Effects of residual superdoses of phytase on growth performance, tibia mineralization, and relative organ weight in ducks fed phosphorus-deficient diets. *Poult. Sci.* **2019**, *98*, 3926–3936. [CrossRef]
27. Green, R.M.; Flamm, S. AGA technical review on the evaluation of liver chemistry tests. *Gastroenterology* **2002**, *123*, 1367–1384. [CrossRef]
28. Sun, H.; Yin, C.Q.; Liu, Q.; Wang, F.; Yuan, C.H. Clinical Significance of Routine Blood Test-Associated Inflammatory Index in Breast Cancer Patients. *Med. Sci. Monit.* **2017**, *23*, 5090–5095. [CrossRef]
29. Ezdini, K.; Ben Salah-Abbès, J.; Belgacem, H.; Mannai, M.; Abbès, S. Lactobacillus paracasei alleviates genotoxicity, oxidative stress status and histopathological damage induced by Fumonisin B1 in BALB/c mice. *Toxicon* **2020**, *185*, 46–56. [CrossRef]
30. Thompson, C.B.; Jakubowski, J.A.; Quinn, P.G.; Deykin, D.; Valeri, C.R. Platelet size and age determine platelet function independently. *Blood* **1984**, *63*, 1372–1375. [CrossRef]
31. Bhandari, N.; He, Q.; Sharma, R.P. Gender-related differences in subacute fumonisin B1 hepatotoxicity in BALB/c mice. *Toxicology* **2001**, *165*, 195–204. [CrossRef]
32. Knodell, R.G.; Ishak, K.G.; Black, W.C.; Chen, T.S.; Craig, R.; Kaplowitz, N.; Kiernan, T.W.; Wollman, J. Formulation and application of a numerical scoring system for assessing histological activity in asymptomatic chronic active hepatitis. *Hepatology* **1981**, *1*, 431–435. [CrossRef]

Article

The Effects of Zearalenone on the Localization and Expression of Reproductive Hormones in the Ovaries of Weaned Gilts

Boyang Wan [1], Xuejun Yuan [2], Weiren Yang [1], Ning Jiao [1], Yang Li [1], Faxiao Liu [1], Mei Liu [1], Zaibin Yang [1], Libo Huang [1,*] and Shuzhen Jiang [1,*]

[1] Department of Animal Sciences and Technology, Shandong Agricultural University, No. 61 Daizong Street, Tai'an 271018, China; sdndwby@163.com (B.W.); wryang@sdau.edu.cn (W.Y.); jiaoning@sdau.edu.cn (N.J.); liyang_cc@yeah.net (Y.L.); 13505388008@163.com (F.L.); liumay@sdau.edu.cn (M.L.); yzb204@163.com (Z.Y.)

[2] Department of Life Sciences, Shandong Agricultural University, No. 61 Daizong Street, Tai'an 271018, China; xjyuan@sdau.edu.cn

*   Correspondence: huanglibo227@126.com (L.H.); shuzhen305@163.com (S.J.)

Abstract: This study aims to investigate the effects of zearalenone (ZEA) on the localizations and expressions of follicle stimulating hormone receptor (FSHR), luteinizing hormone receptor (LHR), gonadotropin releasing hormone (GnRH) and gonadotropin releasing hormone receptor (GnRHR) in the ovaries of weaned gilts. Twenty 42-day-old weaned gilts were randomly allocated into two groups, and treated with a control diet and a ZEA-contaminated diet (ZEA 1.04 mg/kg), respectively. After 7-day adjustment, gilts were fed individually for 35 days and euthanized for blood and ovarian samples collection before morning feeding on the 36th day. Serum hormones of E_2, PRG, FSH, LH and GnRH were determined using radioimmunoassay kits. The ovaries were collected for relative mRNA and protein expression, and immunohistochemical analysis of FSHR, LHR, GnRH and GnRHR. The results revealed that ZEA exposure significantly increased the final vulva area ($p < 0.05$), significantly elevated the serum concentrations of estradiol, follicle stimulating hormone and GnRH ($p < 0.05$), and markedly up-regulated the mRNA and protein expressions of FSHR, LHR, GnRH and GnRHR ($p < 0.05$). Besides, the results of immunohistochemistry showed that the immunoreactive substances of ovarian FSHR, LHR, GnRH and GnRHR in the gilts fed the ZEA-contaminated diet were stronger than the gilts fed the control diet. Our findings indicated that dietary ZEA (1.04 mg/kg) could cause follicular proliferation by interfering with the localization and expression of FSHR, LHR, GnRH and GnRHR, and then affect the follicular development of weaned gilts.

Keywords: zearalenone; gilts; ovary; hormone

Key Contribution: The results showed that ZEA up-regulated the expression of FSHR, LHR, GnRH and GnRHR, and promoted the serum E_2, FSH and GnRH, and thus affected the ovaries of weaned gilts.

Citation: Wan, B.; Yuan, X.; Yang, W.; Jiao, N.; Li, Y.; Liu, F.; Liu, M.; Yang, Z.; Huang, L.; Jiang, S. The Effects of Zearalenone on the Localization and Expression of Reproductive Hormones in the Ovaries of Weaned Gilts. *Toxins* **2021**, *13*, 626. https://doi.org/10.3390/toxins13090626

Received: 12 July 2021
Accepted: 5 September 2021
Published: 7 September 2021

Publisher's Note: MDPI stays neutral with regard to jurisdictional claims in published maps and institutional affiliations.

1. Introduction

Many toxigenic species of *Fusarium* are the main pathogens of cereal plants, causing head blight in wheat and barley and ear rot in maize. Now, there is a lot of evidence that cereals and animals all over the world are polluted by *Fusarium* mycotoxins, especially ZEA. The trade in these commodities may contribute to the spread of this mycotoxin around the world [1,2]. Zearalenone (ZEA) is a kind of exogenous endocrine disruptor mainly produced by *Fusarium fungi* and widely distributed in maize, wheat, barley and other grain crops [3,4]. ZEA-contaminated feed has posed a widespread threat to animals and human beings due to its high stability during storage and heat treatment [5,6]. ZEA can interfere with hormone metabolism by binding with estrogen receptors, and thereby cause reproductive disorders [7,8]. Pigs are the most sensitive animals to ZEA [9]. During the early ovarian development of piglets, the oocytes developed in the follicles are very vulnerable to the intake of nutrients and environmental estrogens. In turn, the intake of nutrients and environmental estrogens affect the level of hormone metabolism, thus changing the ovarian function [10]. Studies have shown that long-term intake of dietary ZEA can result in swelling of reproductive organs, reproductive disruption, abortion and reduction in litter size [11,12].

Reproductive hormones are an indispensable regulator of reproductive organs development and female reproduction. Gonadotropin releasing hormone (GnRH) is a 10 amino acid peptide secreted by the hypothalamus. It is the initial trigger of hypothalamic-pituitary-gonadal axis (HPG) and plays a crucial role in the occurrence of puberty [13]. It can regulate the synthesis and release of follicle stimulating hormone (FSH) and luteinizing hormone (LH), and then promote follicle maturation and gonadal steroid production [14,15]. Gonadotropin releasing hormone receptor (GnRHR) is a G protein coupled receptor. Previous studies have indicated that GnRH combines with GnRHR on the target cell membrane, and activates the intracellular signaling pathway to complete its biological function [16,17]. Follicle stimulating hormone receptor (FSHR) is an important member of HPG axis, which plays a vital role in granulosa cell (GC) proliferation, apoptosis and differentiation, as well as follicle development and ovulation [18,19]. Luteinizing hormone receptor (LHR) is obtained from growing follicles by the combined action of FSH and estradiol (E_2). The expression of LHR in the ovarian cycle changes significantly with the alteration of hormone environment, mainly manifested by the changes in FSH and LH levels. It has been reported that FSH, together with other paracrine factors, regulates the development of the primary follicles toward the preantral and antral stages, and a large number of LHR appear under the FSH stimulation [20]. Previous research has studied the localizations or expressions of GnRHR, FSHR and LHR in the ovaries of rats, ewes and rabbits [21–25]. More and more evidence showed that E_2 and ZEA induced precocious puberty in female rats by increasing the concentrations of serum FSH, LH and GnRH [26–28]. The localization of hormone receptors in different cells can better understand the mechanism of early follicular development induced by ZEA. However, the expressions and localizations of ovarian FSHR, LHR, GnRH and GnRHR in the gilts induced by ZEA have not been elucidated.

Therefore, here, this study was conducted to assess the effects of 1.0 mg/kg ZEA on the localizations and expressions of FSHR, LHR, GnRH and GnRHR in the ovaries of weaned gilts, and we hypothesized that ZEA could change the localizations and expressions of FSHR, LHR, GnRH and GnRHR, and then affect the follicular development of weaned gilts.

2. Results

2.1. Vulva Size

There was no significant difference in the initial vulva area and the final weight between control and the 1.04 mg/kg ZEA treatment ($p > 0.05$, Table 1). The final vulva area, final area/initial area and the final area/final weight of gilts fed the ZEA diet were greater than those of gilts fed the control diet ($p < 0.05$).

2.2. Serum Hormones

Serum E_2, progesterone (PRG), FSH, LH and GnRH levels of gilts are shown in Table 1. Compared with the gilts fed the control diet, gilts fed the 1.04 mg/kg ZEA diet had higher serum levels of E_2, FSH and GnRH ($p < 0.05$). However, ZEA treatment had no effect on serum PRG and LH levels ($p > 0.05$).

2.3. Localizations of FSHR, LHR, GnRH and GnRHR

Immunohistochemical results showed that FSHR immunoreactive substances were mainly localized in the oocytes, GCs and vessel endothelial cells of the ovaries of gilts (Figure 1). Compared with the control group, the positive reactions of FSHR (Figure 1C–J) in oocytes of primordial follicles and granulosa cells of growing follicles in ZEA group were enhanced.

Table 1. Effects of zearalenone (ZEA) on the vulva size and serum hormones of weaned gilts.

Items	Control	ZEA
Vulva size, cm^2		
Initial area	1.39 ± 0.14	1.27 ± 0.12
Final area	2.78 ± 0.09 [b]	4.89 ± 0.18 [a]
Final weight, kg	29.38 ± 2.60	28.66 ± 2.55
Final area/initial area	2.03 ± 0.25 [b]	3.89 ± 0.32 [a]
Final area/final weight, cm^2/kg	0.10 ± 0.01 [b]	0.17 ± 0.02 [a]
Serum hormones		
Estradiol, ng/mL	11.19 ± 0.11 [b]	14.27 ± 0.10 [a]
Progesterone, ng/mL	1.23 ± 0.08	1.19 ± 0.06
Follicle stimulating hormone, mIU/mL	2. 87 ± 0.09 [b]	4.19 ± 0.18 [a]
Luteinizing hormone, mIU/mL	3.39 ± 0.11	3.62 ± 0.21
Gonadotropin releasing hormone, ng/L	12.29 ± 0.27 [b]	17.52 ± 0.48 [a]

Treatments were basal diet supplemented with ZEA at the level of 0 and 1 mg/kg, with analyzed ZEA concentrations of 0 and 1.04 ± 0.03 mg/kg, respectively. Data are mean value ± standard deviation (n = 10). [a,b] Means differ significantly ($p < 0.05$).

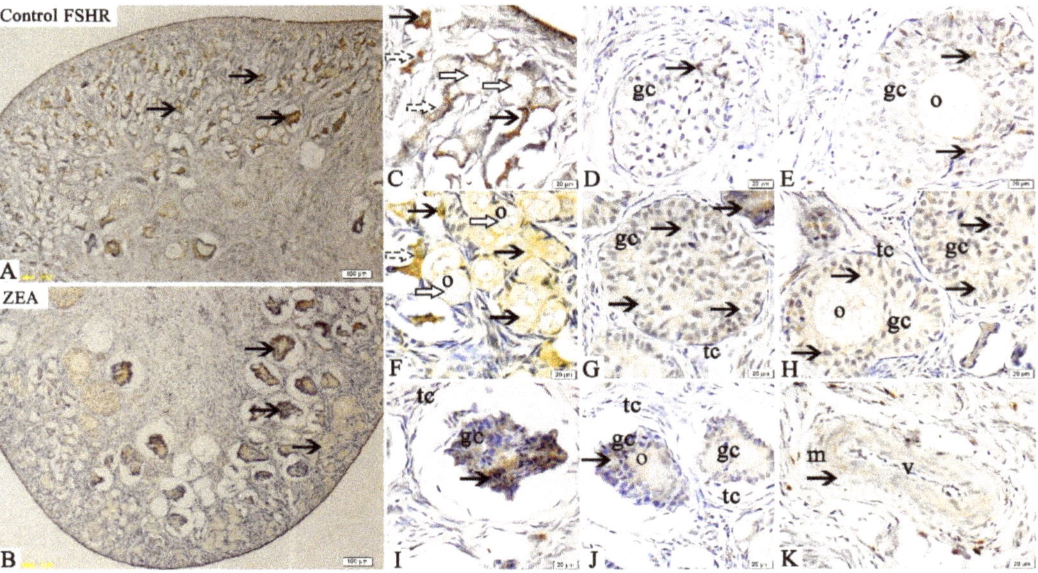

Figure 1. Effects of zearalenone (ZEA) on the follicle stimulating hormone receptor (FSHR) localization in the ovary of weaned gilts. Control (**A,C–E**) and ZEA (**B,F–K**) represent the basal diet with an addition of 0 and 1.0 mg/kg ZEA, and with analyzed ZEA concentrations of 0 and 1.04 ± 0.03 mg/kg, respectively. The o, gc, tc, m and v represent oocyte, granulosa cell, theca cell, smooth muscle and vessel, respectively. The hollow arrows indicate the healthy primordial follicle, the dashed arrows indicate the atresia of the primordial follicle, and the black arrows indicate the FSHR localization in the follicle. Scale bars were 100 μm for A and B, and 20 μm for C–K, respectively (n = 10).

The LHR immunoreactivities were mainly detected in the oocytes, GCs and theca cells of the ovaries in gilts (Figure 2). The overall positive reactions of LHR were weaker than those of FSHR, but the immunoreactive substances of LHR were more obviously observed in the ZEA group compared with the control group (Figure 2C,D,F–H,J). Additionally, the LHR expression was higher in atresia follicles than in healthy growing follicles (Figure 2E,I).

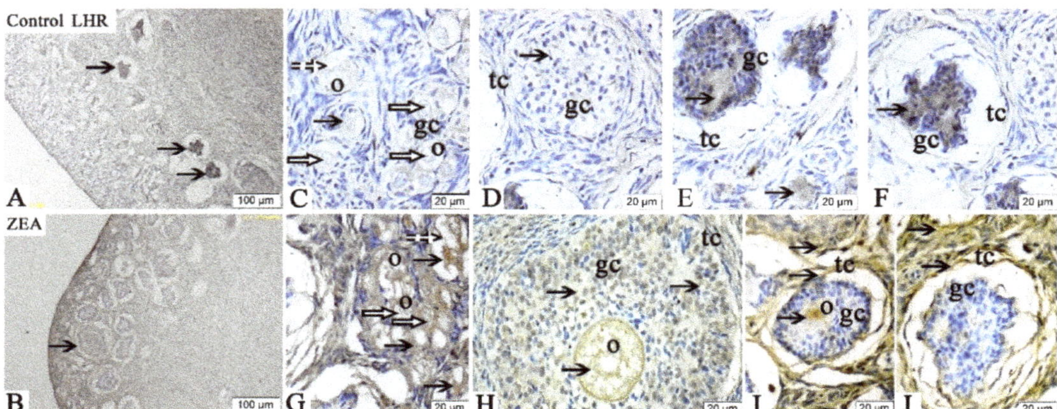

Figure 2. Effects of zearalenone (ZEA) on the luteinizing hormone receptor (LHR) localization in the ovary of weaned gilts. Control (**A**,**C–F**) and ZEA (**B**,**G–J**) represent the basal diet with an addition of 0 and 1.0 mg/kg ZEA, and with analyzed ZEA concentrations of 0 and 1.04 ± 0.03 mg/kg, respectively. The o, gc and tc represent oocyte, granulosa cell and theca cell, respectively. The hollow arrows indicate the healthy primordial follicle, the dashed arrows indicate the atresia of the primordial follicle and the black arrows indicate the LHR localization in the follicle. Scale bars were 100 μm for (**A**,**B**), and 20 μm for (**C–J**), respectively (*n* = 10).

The immunoreactive substances of GnRH were also mainly detected in the oocytes, GCs and vessel endothelial cells of the ovaries in gilts (Figure 3), which showed that the ovary could secrete the GnRH. Zearalenone consumption increased the GnRH expression in oocytes of primordial follicles and GCs of growing follicles, and the staining intensity in the ovary of the ZEA-treated gilts was significantly stronger than that of the control gilts (Figure 3C,D,G). Meanwhile, the GnRH expression in the atresia follicles was higher than that in the growing follicles (Figure 3E,F,H,I).

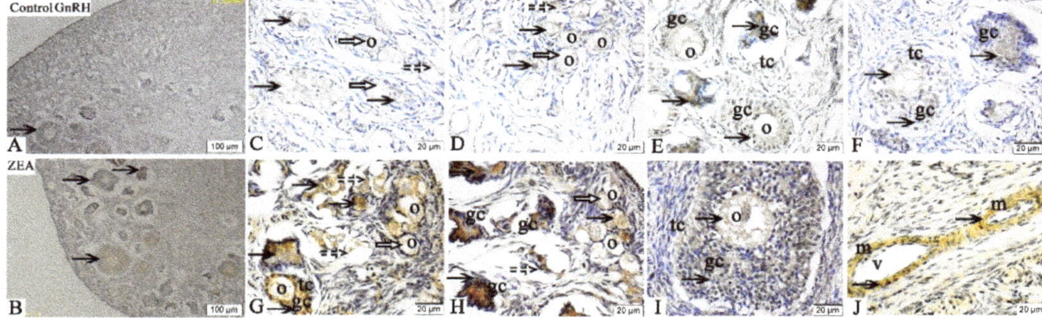

Figure 3. Effects of zearalenone (ZEA) on the gonadotropin releasing hormone (GnRH) localization in the ovary of weaned gilts. Control (**A**,**C–F**) and ZEA (**B**,**G–J**) represent the basal diet with an addition of 0 and 1.0 mg/kg ZEA, and with analyzed ZEA concentrations of 0 and 1.04 ± 0.03 mg/kg, respectively. The o, gc, tc, m and v represent oocyte, granulosa cell, theca cell, smooth muscle and vessel, respectively. The hollow arrows indicate the healthy primordial follicle, the dashed arrows indicate the atresia of the primordial follicle, and the black arrows indicate the GnRH localization in the follicle. Scale bars were 100 μm for (**A**,**B**), and 20 μm for (**C–J**), respectively (*n* = 10).

The GnRHR immunoreactivities were mainly detected in the oocytes, GCs and vessel endothelial cells of the ovaries in gilts (Figure 4). Compared with the control group, the positive reactions of GnRHR (Figure 4C–J) in oocytes of primordial follicles and granulosa cells of growing follicles in ZEA group were enhanced.

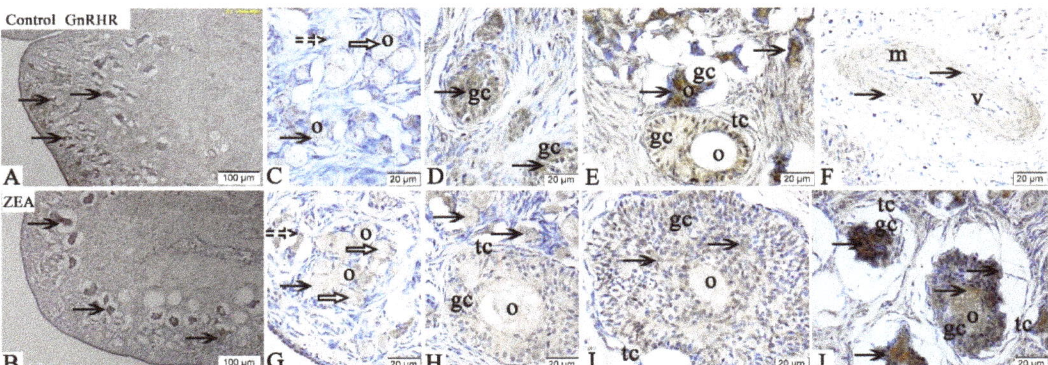

Figure 4. Effects of zearalenone (ZEA) on the gonadotropin releasing hormone receptor (GnRHR) localization in the ovary of weaned gilts. Control (**A,C–F**) and ZEA (**B,G–J**) represent the basal diet with an addition of 0 and 1.0 mg/kg ZEA, and with analyzed ZEA concentrations of 0 and 1.04 ± 0.03 mg/kg, respectively. The o, gc, tc, m and v represent oocyte, granulosa cell, theca cell, smooth muscle and vessel, respectively. The hollow arrows indicate the healthy primordial follicle, the dashed arrows indicate the atresia of the primordial follicle, and the black arrows indicate the GnRHR localization in the follicle. Scale bars were 100 μm for (**A,B**), and 20 μm for (**C–J**), respectively (*n* = 10).

The results of integrated optic density (IOD) of ovarian FSHR, LHR, GnRH and GnRHR in weaned gilts were consistent with the above results of immunochemical analysis results (Table 2). In general, the IOD of FSHR, LHR, GnRH and GnRHR in the ZEA group were higher than those in the control group ($p < 0.05$).

Table 2. Effects of zearalenone (ZEA) on the immunoreactive integrated optic density (IOD) and the mRNA expression of ovarian hormones in weaned gilts.

Items	IOD $\times 10^3$ (*n* = 10)		mRNA Expression (*n* = 6)	
	Control	ZEA	Control	ZEA
Estradiol	5.28 ± 0.36 [b]	11.33 ± 1.53 [a]	1.00 ± 0.12 [b]	3.19 ± 0.12 [a]
Progesterone	4.32 ± 0.35 [b]	4.72 ± 0.16 [a]	1.00 ± 0.09 [b]	1.53 ± 0.04 [a]
Follicle stimulating hormone	2.11 ± 0.26 [b]	4.81 ± 0.34 [a]	1.00 ± 0.07 [b]	2.69 ± 0.21 [a]
Luteinizing hormone	2.89 ± 0.15 [b]	3.13 ± 0.21 [a]	1.00 ± 0.13 [b]	1.96 ± 0.17 [a]

Treatments were basal diet supplemented with ZEA at the level of 0 and 1 mg/kg, with analyzed ZEA concentrations of 0, 1.04 ± 0.03 mg/kg, respectively. Data are mean value ± standard deviation. [a,b] Means differ significantly ($p < 0.05$).

2.4. The mRNA and Protein Expressions

The mRNA and protein expressions of FSHR, LHR, GnRH and GnRHR in ovaries of weaned piglets are shown in Table 2 and Figure 5. The results were consistent with the result of immunohistochemistry analysis. The relative mRNA and protein expressions of FSHR, LHR, GnRH and GnRHR in the ZEA gilts were significantly higher than those in the control gilts ($p < 0.05$).

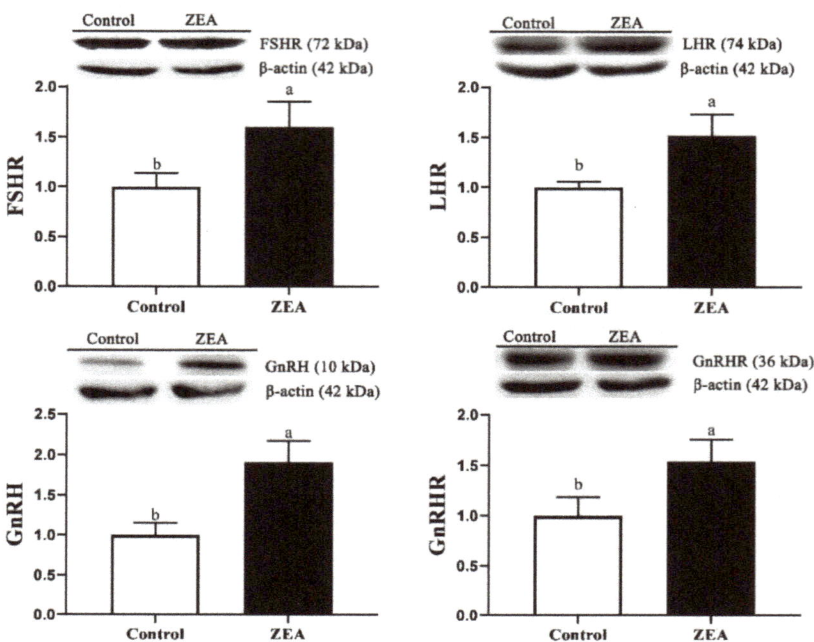

Figure 5. Effects of zearalenone (ZEA) on the protein expressions of follicle stimulating hormone receptor (FSHR), luteinizing hormone receptor (LHR), gonadotropin releasing hormone (GnRH) and gonadotropin releasing hormone receptor (GnRHR) in the ovary of weaned gilts (n = 3). Control and ZEA represent the basal diet with an addition of 0 and 1.0 mg/kg ZEA, and with analyzed ZEA concentrations of 0 and 1.04 ± 0.03 mg/kg, respectively. [a,b] Means differ significantly ($p < 0.05$).

3. Discussion

Various grains and feeds are pervasively polluted by mycotoxins, which causes serious threats to human and animal wellbeing as well as global commercial trade [8]. It is well-documented that as an exogenous estrogen, ZEA can cause hyperestrogenism in pigs [12]. The clinical symptoms of hyperestrogenism were vulva swelling, prolonged oestrus, anal prolapse and high incidence of stillbirth [29–31]. Fu et al. [32] reported that dietary ZEA (1.20 mg/kg) significantly increased vulva width and length of piglets compared to the control group. Similarly, the vulva size of gilts fed with 1.22 mg/kg ZEA increased for a prolonged feeding time [33]. Our previous studies revealed that feeding ZEA-contaminated diets to the gilts (0.96~3.2 mg/kg) could cause vulva swelling and ovarian abnormalities [12,34]. The present results showed that the vulva of weaned gilts fed a ZEA-contaminated diet (1.04 mg/kg) was obviously swollen, suggesting that dietary ZEA may lead to precocious puberty. The most significant change in vulva is mediated by estrogen, which is related to the onset of puberty and the gonadal maturation. In the process of gonadal maturation, follicular development caused the increase in estrogen secretion, which promoted the development of the vulva [35]. It has been reported that endogenous estrogen binds to estrogen receptor subsets and participates in cell proliferation and differentiation through ERK, NF-κB, PI3K/MAPK and other signal transduction pathways [36,37]. However, it is still unclear that whether the mechanisms of ZEA and estrogen leading to vulva swelling is the same, which needs further study.

In the current study, it was successfully observed that dietary ZEA could affect the ovary development by disturbing the reproductive hormones in weaned gilts, which may be highly significant. For early developing animals, mycotoxin poisoning can be determined by examining the changes of serum parameters before clinical symptoms [38]. ZEA and its derivatives can interfere with endocrine, which affects the secretion of steroid hormones [32]. In the present study, we observed that serum E_2 levels increased by ZEA treated, which was similar to the study of Yang et al. [27]. The main role of E_2 is to promote the development of female reproductive organs. The increased serum E_2 concentration was consistent with the appearance of vulva swelling in the present study. The FSH and LH synergistically promotes follicular development and induces the expression of LH receptor on the granulosa cell membrane, which plays a crucial role in the E_2 secretion by the GCs [20]. A previous study indicated that ZEA (5 mg/kg bw) increased the serum FSH level in female rats [27]. Another investigation in female rats also indicated that ZEA (9 and 13.5 mg/kg BW/d) significantly increased serum FSH level in a dose-dependent relationship [28]. Gao et al. [39] demonstrated the promotional effect of ZEA (20 mg/kg) on FSH secretion in maternal rats. Previous study showed that the positive feedback effect (surge) of E_2 significantly increased the expression of proto-oncogene Fos in GnRH cells, whereas short-term removal of negative feedback (ovariectomy) has little and the release of GnRH increased in both states [40]. These changed sexual hormones in weaned gilts are considered to be a marker of premature ovarian failure. These results suggest that ZEA interferes with the normal hormone secretion, which is consistent with our present results. However, there was no significant effect in LH concentration when purified ZEA (1~4 mg/kg) was added to the diets of female rats [26]. Similarly, no significant increase was observed in serum concentration of LH in piglets fed diets containing 596.86 µg/kg ZEA [41], which was consistent with our result. Evans et al. [42] found that E_2 (1.20~7.10 pg/mL) inhibited hypothalamic GnRH secretion in a dose-dependent manner during the time interval between preovulatory luteolysis and gonadotropin surge. It suggests that ZEA interferes with hormone metabolism in a dose-dependent manner. However, it still needs to be further confirmed whether ZEA induced the increased FSH and GnRH by increasing E_2 level first.

The FSHR, LHR, GnRH and GnRHR were mainly located in the ovarian granular layer and follicular membrane [21]. Similarly, in the present study, the FSHR, LHR, GnRH and GnRHR were found to be mainly located in oocytes and granulosa cells of gilts' ovaries. The localization and expression of hormone receptors are very important for a better and clearer understanding of the molecular mechanisms and functions of FSHR, LHR, GnRH and GnRHR. In addition, it is worth noting that FSHR, GnRH and GnRHR showed obvious brown immunoreactive substances in the vascular endothelial cells, suggesting that FSHR, GnRH and GnRHR might play roles in blood vessel generation or blood flow regulation in gilts' ovaries.

FSHR, LHR and GnRHR are all members of the G protein coupled receptor family. Follicle stimulating hormone and LH bind to specific receptors in oocytes, activate intracellular steroidogenic signaling pathway and negatively regulate FSH and LH, thus induced the proliferation and differentiation of ovarian GCs [43–45]. The role of GnRHR in ovarian development is mainly to promote the secretion of FSH and LH through the activity of GnRH [46]. Wei et al. [24] reported that Alarelin immunization could stimulate the production of GnRH antibody, inhibit the expression of GnRHR protein, enhance the expressions of FSHR and LHR protein in the ovary, and increase the secretion of FSH, so as to regulate the development of ovary and follicle in ewes. Other, similar research indicated that FSH could promote the development of ovine oocytes, reduce the apoptosis rate, increase the mRNA expressions of FSHR, LHR and GnRHR, and protein expressions of FSHR and GnRHR [24]. ZEA can compete with endogenous estrogen to bind the estrogen receptor, so as to interfere with ovarian gonadal hormone release, which directly affects the expression and function of hormone receptor in female animals, especially animals during reproductive cycle [47]. In our study, the mRNA and protein expressions of FSHR, LHR and GnRHR in the oocytes and GCs of piglets significantly increased by 1.04 mg/kg ZEA treated. In addition, the IOD of FSHR, LHR, GnRH and GnRHR in the ZEA group were higher than those in the control group. To the best of our knowledge, the study is the first to suggest in vivo that dietary ZEA (1.04 mg/kg) can affect the expressions of FSHR, LHR, GnRH and GnRHR, as observed by immunohistochemistry, in the ovary of weaned gilts. Our immunohistochemistry results strongly indicated that hormone receptors play a crucial role in ZEA-induced ovarian dysplasia. Therefore, we speculate that the mRNA and protein expression of GnRHR in the primordial follicles and GCs of gilts are significantly correlated with the protein expression of FSHR and LHR. However, the molecular mechanism needs to be further verified.

In our study, it can be clear that dietary ZEA significantly increased the mRNA and protein expression of GnRH in gilts in a dose-dependent relationship, which was consistent with the previous studies in Kriszet et al. [48] and Yang et al. [27]. GnRH is an important neuropeptide, which can enter pituitary via HPG axis, stimulate the synthesis of LH and FSH, and then regulate the development and function of the gonad [49–51]. Therefore, we speculate that ZEA first continuously activates the hypothalamus, upregulates GnRH expression, then induces pituitary to release gonadotropin into serum, and finally leads to vaginal swelling and ovarian weight increase. However, the mechanism of ZEA regulating hypothalamus-pituitary-ovary axis remains to be further confirmed.

4. Conclusions

In conclusion, ZEA (1.04 mg/kg) can upregulate the expressions of FSHR, LHR, GnRH and GnRHR in ovaries of weaned gilts, promote the secretion of E_2, FSH and GnRH, and thereby accelerate vulva swelling and follicular hyperplasia. Therefore, ZEA regulates the development of the vulva and ovary by disordering with the reproductive hormone pathway in gilts. This study laid a foundation for finding sensitive indexes and prevention for targets of reproductive disorders caused by ZEA. However, the mechanisms of ZEA-induced vulva swelling and ovarian development by regulating the hypothalamus-pituitary-ovary axis remains to be further studied.

5. Materials and Methods

All agreements used were complied with the Guide for the Care and Use of Laboratory Animals and approved by the Committee on the Ethics (Approval Number: S20180058) of Shandong Agricultural University (Tai'an, China).

5.1. Animals, Treatments and Feeding Management

Twenty healthy weaned gilts (Duroc × Landrace × Yorkshire) at 42-day with an average BW of 12.84 ± 0.26 kg were selected and randomly allocated into two treatments, with 10 replicates per treatment. Gilts were housed individually in stainless-steel cages (0.48 m^2) fitted with plastic slatted floors, feed troughs and nipple drinkers for 35-day test period after 7-day adjustment at the Animal Research Station of Shandong Agricultural University (Tai'an, Shandong, China). During the experimental period, gilts were fed a basal diet (control group) or a ZEA-contaminated diet (the basal diet supplemented with 1.0 mg/kg ZEA). Zearalenone levels used in the present study were based on our previous investigations in Shandong Province of China from 2007 to 2020 and recent literature [12,52–54]. The basal diet (Table 3) used in the present study was prepared according to the NRC (2012) [55]. Diets were completed in one batch, sampled and stored in covered containers before feeding. Before the test, the house was cleaned and disinfected. During the first week of the experiment, the room temperature was set to 30 °C, and then maintained between 26 and 28 °C. The relative humidity was approximately 65%.

Table 3. Ingredients and compositions of the basal diet (as fed basis).

Ingredients	Content (%)	Nutrients	Analyzed Values (%)
Corn	53.00	Metabolizable energy, MJ/kg	13.22
Wheat middling	5.00	Crude protein	19.40
Whey powder	6.50	Calcium	0.84
Soybean oil	2.50	Total phosphorus	0.73
Soybean meal	24.76	Lysine	1.36
Fish meal	5.50	Methionine	0.46
L-Lysine HCl	0.30	Sulfur amino acid	0.79
DL-methionine	0.10	Threonine	0.90
L-threonine	0.04	Tryptophan	0.25
Calcium phosphate	0.80		
Limestone, Pulverized	0.30		
Sodium chloride	0.20		
Premix [1]	1.00		
Total	100		

[1] Supplied per kilogram of diet: vitamin A, 3300 IU; vitamin D_3, 330 IU; vitamin E, 24 IU; vitamin K_3, 0.75 mg; vitamin B_1, 1.50 mg; vitamin B_2, 5.25 mg; vitamin B_6, 2.25 mg; vitamin B_{12}, 0.02625 mg; pantothenic acid, 15.00 mg; niacin, 22.5 mg; biotin, 0.075 mg; folic acid, 0.45 mg; Mn ($MnSO_4 \cdot H_2O$), 6.00 mg; Fe ($FeSO_4 \cdot H_2O$), 150 mg; Zn ($ZnSO_4 \cdot H_2O$), 150 mg; Cu ($CuSO_4 \cdot 5H_2O$), 9.00 mg; I (KIO_3), 0.21 mg; Se (Na_2SeO_3), 0.45 mg.

The nutrients were analyzed according to AOAC (2012) [56]. Mycotoxins were detected by the Qingdao Entry–Exit Inspection and Quarantine Bureau according to the methods of Liu et al. [52], and the minimum detection concentration for ZEA, aflatoxin, fumonisin and deoxynivalenol were 0.01 mg/kg, 1.0 µg/kg, 0.1 mg/kg and 0.05 mg/kg, respectively. The analyzed ZEA contents in the basal diet and ZEA-contaminated diet were <0.01 and 1.04 ± 0.03 mg/kg, respectively, and no other toxins were detected or below the minimum detection concentration.

5.2. Vulva Measurement

The length and width of vulva were measured with Vernier caliper every three days to determine the estrogenic effect of ZEA, and the vulva area was approximately calculated as a diamond shape [(vulva length × vulva width)/2] as described by Jiang et al. [57] and Zhou et al. [12].

5.3. Serum and Ovary Samples Collection

Gilts were fasted for 12 h on the last day of the feeding trial, and then about 10 mL blood samples were taken from jugular vein into tubes without anticoagulant. The blood was incubated at 37 °C for 2 h and then centrifuged at $3000 \times g$ for 15 min to obtain the serum, followed by immediately stored in 1.5 mL Eppendorf tubes at −20 °C for the analysis of E_2, PRG, FSH, LH and GnRH.

Two ovarian samples were isolated from each pig under sterile conditions after euthanasia by electrocution (head only, 110 V, 60 Hz). One of each pair was stored at −80 °C for subsequent analysis of gene and protein expressions of FSHR, LHR, GnRH and GnRHR, and the other was promptly fixed in Bouin's solution for 24 to 48 h for immunohistochemical analysis. Six ovaries were randomly selected for mRNA expression analysis and three ovaries for protein expression analysis.

5.4. Serum Hormone Measurement

Serum levels of E_2, PRG, FSH, LH and GnRH were determined using radioimmunoassay kits (Nanjing Jiancheng Bioengineering Institute, Nanjing, China) according to method previously described by Jiang et al. [57].

5.5. Immunohistochemistry (IHC)

Sections were processed in accordance with the standard IHC protocols. After dewaxing, rehydration and antigen retrieval was performed by microwaving for 20 min at full power in sodium citrate buffer (0.01 mol/L, pH = 6.0). The sections were subsequently treated with 10% hydrogen peroxide (H_2O_2) for 1.5 h to deactivate endogenous peroxidase activity and incubated in 10% normal goat serum (ZSGB-BIO, Beijing, China) for 1 h to block nonspecific binding.

The immunohistochemical analysis was performed using a commercial kit (Polink-2 plus® Polymer HRP Detection system for rabbit primary antibody, PV-9001, ZSGB-BIO, Beijing, China) according to the manufacturer's instructions. Briefly, after washing with phosphate-buffered saline (PBS), the above prepared sections were incubated with anti-FSHR (1:100, bs-0895R, BIOSS, Beijing, China), anti-LHR (1:100, bs-6431R, BIOSS, Beijing, China), anti-GnRH (1:200, 26950-1-AP, Proteintech, Wuhan, China) and anti-GnRHR (1:200, 19950-1-AP, Proteintech, Wuhan China) at 4 °C. The sections were washed with PBS the following day and were subsequently incubated in polymer helper for 1 h at 37 °C followed by Polink-2 plus polymer HRP antirabbit at 37 °C for 1 h. After this incubation, the sections were washed with PBS, followed by immersion in diaminobenzidine tetrachloride (DAB) using a kit (DAB kit, TIANGEN PA110, Beijing, China) for 1~3 min to detect immunostaining. The sections were then dehydrated, sealed in clear resin, mounted and observed microscopically for the localization of immunoreactive substances using a bright field of view.

5.6. Integrated Optical Density Measurement

The FSHR, LHR, GnRH and GnRHR labeling was examined by a microscope (Nikon ELIPSE 80i, Tokyo, Japan). Three stained sections of each sample were randomly selected, and three visual fields of each stained section were randomly selected for observation and photography. To estimate the amount of cell staining, the pictures were analyzed using an image analysis software (Image Pro-Plus 6.0, Media Cybernetics, Silver Spring, MD, USA) [58]. Total cross-sectional IOD was acquired, which was used to compare the FSHR, LHR, GnRHR and GnRH staining intensity between control and ZEA treatments. The IOD of each sample is the average of three stained sections.

5.7. The mRNA Expression Using Quantitative Real-Time Polymerase Chain Reaction (qRT-PCR)

Total RNA was extracted from the ovaries of gilts using RNAiso Plus (Applied TaKaRa, DaLian, China) according to manufacturer's instructions and the literature of Song et al. [53] and Zhou et al. [59], and the purity and concentration of the RNA were evaluated by Eppendorf Biophotometer (DS-11, Denovix, Wilmington, DE, USA) at an absorbance ratio of 260/280 nm (a range of 1.8~2.0 indicates a pure RNA sample). The RNA integrity was verified by agarose gel electrophoresis. Total RNA was reverse transcribed to cDNA using a Reverse Transcription System kit (PrimeScript™ RT Master Mix, RR036A, Applied TaKaRa, DaLian, China).

For qRT-PCR, the total volume of the PCR reaction mixture was 20 μL, which contained 10 μL of SYBR Premix Ex Taq-TIi RNaseH Plus (code: RR420A, Lot: AK7502; Applied TaKaRa, DaLian, China), 0.4 μL of both forward and reverse primers, 0.4 μL DyeII, and 2 μL cDNA (<100 ng). The optimized qRT-PCR protocol included an initial denaturation step at 95 °C for 30 s, followed by 43 cycles at 95 °C for 5 s, 60 °C for 34 s, 95 °C for 15 s and 60 °C for 60 s, with a final step at 95 °C for 15 s. The qRT-PCR reactions were conducted in an ABI 7500 Real Time PCR System (Applied Biosystems, Foster City, CA, USA). The relative mRNA was expressed and calculated as equal to $2^{-\Delta\Delta CT}$ [60]. The analysis was repeated three times for each sample. The primer sequences and production lengths are presented in Table 4.

Table 4. Primers sequences of glycerol triphosphate dehydrogenase (GAPDH), follicle stimulating hormone receptor (FSHR), luteinizing hormone receptor (LHR), gonadotropin releasing hormone (GnRH) and gonadotropin releasing hormone receptor (GnRHR).

Target Genes	Primer Sequence (5′ to 3′)	Product Size	Accession No.
GADPH	F: ATGGTGAAGGTCGGAGTGAA R: CGTGGGTGGAATCATACTGG	154	NM_001206359.1
FSHR	F: ATGTCCTTGCTCCTGGTGTC R: GGTCCCCAAATCCAGAAAAT	213	NM_214386.1
LHR	F: GAAAGCACAGCAAGGAGACC R: ACATGAGGAAACGAGGCACT	282	NM_214449.1
GnRH	F: AGCCAACACTGGTCCTATCGATTG R: GTCTTCTGCCCAGTTTCCTCTTCA	206	NM_214274.1
GnRHR	F: AGCCAACCTGTTGGAGACTCTGAT R: AGCTGAGGACTTTGCAGAGGAACT	101	NM_214273

5.8. Western Blotting

Ovarian protein was extracted according to the lysate instructions (Beyotime, Shanghai, China), and concentrations were determined using a bicinchoninic acid (BCA) protein assay kit (Beyotime, Shanghai, China) with protein content of each sample being adjusted to 55 μg per sample. The proteins were separated by electrophoresis on polyacrylamide gels and were transferred onto immobilon-p transfer membranes (Solarbio, Beijing, China). The membranes were incubated in 10% skimmed milk for 2 h, washed three times with Tris-buffered saline containing Tween (TBST), and then incubated with primary antibodies: anti-FSHR (1:500, bs-0895R, BIOSS, Beijing, China), anti-LHR (1:500, bs-6431R, BIOSS, Beijing, China), anti-GnRH (1:500, 26950-1-AP, Proteintech, Wuhan, China) and anti-GnRHR (1:500, 19950-1-AP, Proteintech, Wuhan, China), diluted with primary antibody dilution buffer (Beyotime, Shanghai, China), at 4 °C overnight. After washing with TBST, the membranes were incubated with antirabbit IgG (1:1000, Beyotime, Shanghai, China), which were diluted by secondary antibody dilution buffer (Beyotime, Shanghai, China) at 37 °C for 2 h, immersed in a high-sensitivity luminescence reagent (BeyoECL Plus, Beyotime, Shanghai, China), exposed to film using FusionCapt Advance FX7 (Beijing Oriental Science and Technology Development Co. Ltd., Beijing, China), and then quantified using Image software (Image Pro-Plus 6.0, Media Cybernetics, Silver Spring, MD, USA).

5.9. Statistical Analysis

Individual piglet was taken as the experimental unit. To determine the difference between control and the ZEA treatment, the data were statistically analyzed using a two-sample pairwise t-test with SAS 9.2 statistical software (SAS Institute Inc., Cary, NC, USA). All data are expressed as the mean ± standard deviation (SD). The difference was considered significant when $p < 0.05$.

Author Contributions: Conceptualization, L.H. and S.J.; data curation, B.W., X.Y. and M.L.; formal analysis, F.L., L.H. and S.J.; funding acquisition, W.Y., Z.Y., L.H. and S.J.; methodology, W.Y., Z.Y., L.H. and S.J.; writing—original draft, B.W.; writing—review and editing, N.J., Y.L. and S.J. All authors have read and agreed to the published version of the manuscript.

Funding: This research was financed in part by the Natural Science Foundation of Shandong Province (grant no. ZR2019MC038); Major Innovative Projects in Shandong Province of Research and application of environment-friendly feed and the critical technologies for pigs and poultry without antibiotic (grant no. 2019JZZY020609); Founds of Shandong Agriculture Research System in Shandong Province (grant no. SDAIT-08-04).

Institutional Review Board Statement: The authors confirm that the ethical policies of the journal, as noted on the journal's author guidelines page, have been adhered to. All animal care and experimental procedures were in accordance with the guidelines for the care and use of laboratory animals prescribed by the Animal Nutrition Research Institute of Shandong Agricultural University and the Ministry of Agriculture of China.

Informed Consent Statement: Not applicable.

Data Availability Statement: Not applicable.

Acknowledgments: Financial supports of the natural science fund, the major innovation fund and the pig industry fund of Shandong Province.

Conflicts of Interest: We certify that there are no conflicts of interest with any financial organization regarding the material discussed in the manuscript.

References

1. Gautier, C.; Pinson-Gadais, L.; Richard-Forget, F. Fusarium Mycotoxins Enniatins: An updated review of their occurrence, the producing fusarium species, and the abiotic determinants of their accumulation in crop harvests. *J. Agric. Food Chem.* **2020**, *68*, 4788–4798. [CrossRef] [PubMed]
2. Placinta, C.M.; D'Mello, J.P.F.; Macdonald, A.M.C. A review of worldwide contamination of cereal grains and animal feed with Fusarium mycotoxins. *Anim. Feed Sci. Technol.* **1999**, *78*, 21–37. [CrossRef]
3. Rai, A.; Das, M.; Tripathi, A. Occurrence and toxicity of a fusarium mycotoxin, zearalenone. *Crit. Rev. Food Sci. Nutr.* **2020**, *60*, 2710–2729. [CrossRef] [PubMed]
4. Yang, D.; Jiang, X.; Sun, J.; Li, X.; Li, X.; Jiao, R.; Peng, Z.; Li, Y.; Bai, W. Toxic effects of zearalenone on gametogenesis and embryonic development: A molecular point of review. *Food Chem. Toxicol.* **2018**, *119*, 24–30. [CrossRef]
5. Rogowska, A.; Pomastowski, P.; Sagandykova, G.; Buszewski, B. Zearalenone and its metabolites: Effect on human health, metabolism and neutralisation methods. *Toxicon* **2019**, *162*, 46–56. [CrossRef]
6. Zinedine, A.; Soriano, J.M.; Molto, J.C.; Manes, J. Review on the toxicity, occurrence, metabolism, detoxification, regulations and intake of zearalenone: An oestrogenic mycotoxin. *Food Chem. Toxicol.* **2007**, *45*, 1–18. [CrossRef]
7. Ahmad, B.; Shrivastava, V.K.; Saleh, R.; Henkel, R.; Agarwal, A. Protective effects of saffron against zearalenone-induced alterations in reproductive hormones in female mice (Mus musculus). *Clin. Exp. Reprod. Med.* **2018**, *45*, 163–169. [CrossRef]
8. Mahato, D.K.; Devi, S.; Pandhi, S.; Sharma, B.; Maurya, K.K.; Mishra, S.; Dhawan, K.; Selvakumar, R.; Kamle, M.; Mishra, A.K.; et al. Occurrence, impact on agriculture, human health, and management strategies of zearalenone in food and feed: A review. *Toxins (Basel)* **2021**, *13*, 92. [CrossRef]
9. Pistol, G.C.; Braicu, C.; Motiu, M.; Gras, M.A.; Marin, D.E.; Stancu, M.; Calin, L.; Israel-Roming, F.; Berindan-Neagoe, I.; Taranu, I. Zearalenone mycotoxin affects immune mediators, MAPK signalling molecules, nuclear receptors and genome-wide gene expression in pig spleen. *PLoS ONE* **2015**, *10*, e0127503. [CrossRef]
10. Jarrett, S.; Ashworth, C.J. The role of dietary fibre in pig production, with a particular emphasis on reproduction. *J. Anim. Sci. Biotechnol.* **2018**, *9*, 59. [CrossRef]
11. Ropejko, K.; Twaruzek, M. Zearalenone and its metabolites-general overview, occurrence, and toxicity. *Toxins* **2021**, *13*, 35. [CrossRef]
12. Zhou, M.; Yang, L.; Chen, Y.; Sun, T.; Wang, N.; Chen, X.; Yang, Z.; Ge, J.; Jiang, S. Comparative study of stress response, growth and development of uteri in post-weaning gilts challenged with zearalenone and estradiol benzoate. *J. Anim. Physiol. Anim. Nutr. (Berl.)* **2019**, *103*, 1885–1894. [CrossRef]
13. Huang, W.; Yao, B.; Sun, L.; Pu, R.; Wang, L.; Zhang, R. Immunohistochemical and in situ hybridization studies of gonadotropin releasing hormone (GnRH) and its receptor in rat digestive tract. *Life Sci.* **2001**, *68*, 1727–1734. [CrossRef]
14. Haen, S.M.; Heinonen, M.; Kauffold, J.; Heikinheimo, M.; Hoving, L.L.; Soede, N.M.; Peltoniemi, O.A.T. GnRH-agonist deslorelin implant alters the progesterone release pattern during early pregnancy in gilts. *Reprod. Domest. Anim.* **2019**, *54*, 464–472. [CrossRef]
15. Kadokawa, H.; Pandey, K.; Nahar, A.; Nakamura, U.; Rudolf, F.O. Gonadotropin-releasing hormone (GnRH) receptors of cattle aggregate on the surface of gonadotrophs and are increased by elevated GnRH concentrations. *Anim. Reprod. Sci.* **2014**, *150*, 84–95. [CrossRef] [PubMed]
16. Sakai, T.; Yamamoto, T.; Matsubara, S.; Kawada, T.; Satake, H. Invertebrate gonadotropin-releasing hormone receptor signaling and its relevant biological actions. *Int. J. Mol. Sci.* **2020**, *21*, 8544. [CrossRef] [PubMed]
17. Schang, A.L.; Querat, B.; Simon, V.; Garrel, G.; Bleux, C.; Counis, R.; Cohen-Tannoudji, J.; Laverriere, J.N. Mechanisms underlying the tissue-specific and regulated activity of the Gnrhr promoter in mammals. *Front. Endocrinol. (Lausanne)* **2012**, *3*, 162. [CrossRef]
18. Wang, N.; Wu, W.; Pan, J.; Long, M. Detoxification strategies for zearalenone using microorganisms: A review. *Microorganisms* **2019**, *7*, 208. [CrossRef] [PubMed]
19. Xu, J.; Gao, X.; Li, X.; Ye, Q.; Jebessa, E.; Abdalla, B.A.; Nie, Q. Molecular characterization, expression profile of the FSHR gene and its association with egg production traits in muscovy duck. *J. Genet.* **2017**, *96*, 341–351. [CrossRef]
20. Menon, K.M.; Nair, A.K.; Wang, L.; Peegel, H. Regulation of luteinizing hormone receptor mRNA expression by a specific RNA binding protein in the ovary. *Mol. Cell Endocrinol.* **2007**, *260-262*, 109–116. [CrossRef] [PubMed]

21. Chen, H.; Cui, Y.; Yu, S. Expression and localisation of FSHR, GHR and LHR in different tissues and reproductive organs of female yaks. *Folia Morphol. (Warsz)* **2018**, *77*, 301–309. [CrossRef]

22. Lan, R.X.; Liu, F.; He, Z.B.; Chen, C.; Liu, S.J.; Shi, Y.; Liu, Y.L.; Yoshimura, Y.; Zhang, M. Immunolocalization of GnRHRI, gonadotropin receptors, PGR, and PGRMCI during follicular development in the rabbit ovary. *Theriogenology* **2014**, *81*, 1139–1147. [CrossRef]

23. Sengupta, A.; Chakrabarti, N.; Sridaran, R. Presence of immunoreactive gonadotropin releasing hormone (GnRH) and its receptor (GnRHR) in rat ovary during pregnancy. *Mol. Reprod. Dev.* **2008**, *75*, 1031–1044. [CrossRef]

24. Wei, S.; Chen, S.; Gong, Z.; Ouyang, X.; An, L.; Xie, K.; Dong, J.; Wei, M. Alarelin active immunization influences expression levels of GnRHR, FSHR and LHR proteins in the ovary and enhances follicular development in ewes. *Anim. Sci. J.* **2013**, *84*, 466–475. [CrossRef] [PubMed]

25. Wen, X.; Xie, J.; Zhou, L.; Fan, Y.; Yu, B.; Chen, Q.; Fu, Y.; Yan, Z.; Guo, H.; Lyu, Q.; et al. The role of combining medroxyprogesterone 17-acetate with human menopausal gonadotropin in mouse ovarian follicular development. *Sci. Rep.* **2018**, *8*, 4439. [CrossRef] [PubMed]

26. Collins, T.F.; Sprando, R.L.; Black, T.N.; Olejnik, N.; Eppley, R.M.; Alam, H.Z.; Rorie, J.; Ruggles, D.I. Effects of zearalenone on in utero development in rats. *Food Chem. Toxicol.* **2006**, *44*, 1455–1465. [CrossRef] [PubMed]

27. Yang, R.; Wang, Y.M.; Zhang, L.; Zhao, Z.M.; Zhao, J.; Peng, S.Q. Prepubertal exposure to an oestrogenic mycotoxin zearalenone induces central precocious puberty in immature female rats through the mechanism of premature activation of hypothalamic kisspeptin-GPR54 signaling. *Mol. Cell Endocrinol.* **2016**, *437*, 62–74. [CrossRef]

28. Zhang, Y.; Jia, Z.; Yin, S.; Shan, A.; Gao, R.; Qu, Z.; Liu, M.; Nie, S. Toxic effects of maternal zearalenone exposure on uterine capacity and fetal development in gestation rats. *Reprod. Sci.* **2014**, *21*, 743–753. [CrossRef]

29. Gajecka, M.; Rybarczyk, L.; Jakimiuk, E.; Zielonka, L.; Obremski, K.; Zwierzchowski, W.; Gajecki, M. The effect of experimental long-term exposure to low-dose zearalenone on uterine histology in sexually immature gilts. *Exp. Toxicol. Pathol.* **2012**, *64*, 537–542. [CrossRef] [PubMed]

30. Tian, Y.; Zhang, M.Y.; Zhao, A.H.; Kong, L.; Wang, J.J.; Shen, W.; Li, L. Single-cell transcriptomic profiling provides insights into the toxic effects of zearalenone exposure on primordial follicle assembly. *Theranostics* **2021**, *11*, 5197–5213. [CrossRef]

31. Wang, J.P.; Chi, F.; Kim, I.H. Effects of montmorillonite clay on growth performance, nutrient digestibility, vulva size, faecal microflora, and oxidative stress in weaning gilts challenged with zearalenone. *Anim. Feed Sci. Technol.* **2012**, *178*, 158–166. [CrossRef]

32. Fu, G.; Ma, J.; Wang, L.; Yang, X.; Liu, J.; Zhao, X. Effect of degradation of zearalenone-contaminated feed by *Bacillus licheniformis* CK1 on postweaning female oiglets. *Toxins* **2016**, *8*, 300. [CrossRef] [PubMed]

33. Su, Y.; Sun, Y.; Ju, D.; Chang, S.; Shi, B.; Shan, A. The detoxification effect of vitamin C on zearalenone toxicity in piglets. *Ecotoxicol. Environ. Saf.* **2018**, *158*, 284–292. [CrossRef] [PubMed]

34. Jiang, S.Z.; Yang, Z.B.; Yang, W.R.; Gao, J.; Liu, F.X.; Broomhead, J.; Chi, F. Effects of purified zearalenone on growth performance, organ size, serum metabolites, and oxidative stress in postweaning gilts. *J. Anim. Sci.* **2011**, *89*, 3008–3015. [CrossRef] [PubMed]

35. Farage, M.; Maibach, H. Lifetime changes in the vulva and vagina. *Arch. Gynecol. Obstet.* **2006**, *273*, 195–202. [CrossRef]

36. Kiyama, R.; Wada-Kiyama, Y. Estrogenic endocrine disruptors: Molecular mechanisms of action. *Environ. Int.* **2015**, *83*, 11–40. [CrossRef]

37. Wang, C.; Xu, C.X.; Bu, Y.; Bottum, K.M.; Tischkau, S.A. Beta-naphthoflavone (DB06732) mediates estrogen receptor-positive breast cancer cell cycle arrest through AhR-dependent regulation of PI3K/AKT and MAPK/ERK signaling. *Carcinogenesis* **2014**, *35*, 703–713. [CrossRef] [PubMed]

38. Oguz, H.; Kececi, T.; Birdane, Y.O.; Onder, F.; Kurtoglu, V. Effect of clinoptilolite on serum biochemical and haematological characters of broiler chickens during aflatoxicosis. *Res. Vet. Sci.* **2000**, *69*, 89–93. [CrossRef]

39. Gao, X.; Sun, L.; Zhang, N.; Li, C.; Zhang, J.; Xiao, Z.; Qi, D. Gestational zearalenone exposure causes reproductive and developmental toxicity in pregnant rats and female offspring. *Toxins* **2017**, *9*, 21. [CrossRef]

40. Moenter, S.M.; Karsch, F.J.; Lehman, M.N. Fos expression during the estradiol-induced gonadotropin-releasing hormone (GnRH) surge of the ewe: Induction in GnRH and other neurons. *Endocrinology* **1993**, *133*, 896–903. [CrossRef]

41. Shi, D.; Zhou, J.; Zhao, L.; Rong, X.; Fan, Y.; Hamid, H.; Li, W.; Ji, C.; Ma, Q. Alleviation of mycotoxin biodegradation agent on zearalenone and deoxynivalenol toxicosis in immature gilts. *J. Anim. Sci. Biotechnol.* **2018**, *9*, 42. [CrossRef] [PubMed]

42. Evans, N.P.; Dahl, G.E.; Glover, B.H.; Karsch, F.J. Central regulation of pulsatile gonadotropin-releasing hormone (GnRH) secretion by estradiol during the period leading up to the preovulatory GnRH surge in the ewe. *Endocrinology* **1994**, *134*, 1806–1811. [CrossRef]

43. Lai, L.; Shen, X.; Liang, H.; Deng, Y.; Gong, Z.; Wei, S. Determine the role of FSH receptor binding inhibitor in regulating ovarian follicles development and expression of FSHR and ERalpha in mice. *BioMed Res. Int.* **2018**, *2018*, 5032875. [CrossRef] [PubMed]

44. Milazzotto, M.P.; Rahal, P.; Nichi, M.; Miranda-Neto, T.; Teixeira, L.A.; Ferraz, J.B.S.; Eler, J.P.; Campagnari, F.; Garcia, J.F. New molecular variants of hypothalamus–pituitary–gonad axis genes and their association with early puberty phenotype in Bos taurus indicus (Nellore). *Livest. Sci.* **2008**, *114*, 274–279. [CrossRef]

45. Zhang, Q.; Madden, N.E.; Wong, A.S.T.; Chow, B.K.C.; Lee, L.T.O. The role of endocrine G protein-coupled receptors in ovarian cancer progression. *Front. Endocrinol. (Lausanne)* **2017**, *8*, 66. [CrossRef]

46. Suocheng, W.; Zhuandi, G.; Li, S.; Haoqin, L.; Luju, L.; Yingying, D. Maturation rates of oocytes and levels of FSHR, LHR and GnRHR of COCs response to FSH concentrations in IVM media for sheep. *J. Appl. Biomed.* **2017**, *15*, 180–186. [CrossRef]

47. Silva, J.R.; van den Hurk, R.; de Matos, M.H.; dos Santos, R.R.; Pessoa, C.; de Moraes, M.O.; de Figueiredo, J.R. Influences of FSH and EGF on primordial follicles during in vitro culture of caprine ovarian cortical tissue. *Theriogenology* **2004**, *61*, 1691–1704. [CrossRef]

48. Kriszt, R.; Winkler, Z.; Polyak, A.; Kuti, D.; Molnar, C.; Hrabovszky, E.; Kallo, I.; Szoke, Z.; Ferenczi, S.; Kovacs, K.J. Xenoestrogens ethinyl estradiol and zearalenone cause precocious puberty in female rats via central isspeptin signaling. *Endocrinology* **2015**, *156*, 3996–4007. [CrossRef]

49. Knox, R.V.; Webel, S.K.; Swanson, M.; Johnston, M.E.; Kraeling, R.R. Effects of estrus synchronization using Matrix® followed by treatment with the GnRH agonist triptorelin to control ovulation in mature gilts. *Anim. Reprod. Sci.* **2017**, *185*, 66–74. [CrossRef] [PubMed]

50. Qin, X.; Xiao, Y.; Ye, C.; Jia, J.; Liu, X.; Liang, H.; Zou, G.; Hu, G. Pituitary Action of E2 in Prepubertal grass carp: Receptor specificity and signal transduction for luteinizing hormone and follicle-stimulating hormone regulation. *Front. Endocrinol. (Lausanne)* **2018**, *9*, 308. [CrossRef]

51. Terasawa, E. Neuroestradiol in regulation of GnRH release. *Horm. Behav.* **2018**, *104*, 138–145. [CrossRef]

52. Liu, X.; Xu, C.; Yang, Z.; Yang, W.; Huang, L.; Wang, S.; Liu, F.; Liu, M.; Wang, Y.; Jiang, S. Effects of dietary zearalenone exposure on the growth performance, small intestine disaccharidase, and antioxidant activities of weaned gilts. *Animals* **2020**, *10*, 2157. [CrossRef] [PubMed]

53. Song, T.; Liu, X.; Yuan, X.; Yang, W.; Liu, F.; Hou, Y.; Huang, L.; Jiang, S. Dose-effect of zearalenone on the localization and expression of growth hormone, growth hormone receptor, and heat shock protein 70 in the ovaries of post-weaning gilts. *Front. Vet. Sci.* **2021**, *8*, 629006. [CrossRef] [PubMed]

54. Chen, X.X.; Yang, C.W.; Huang, L.B.; Niu, Q.S.; Jiang, S.Z.; Chi, F. Zearalenone altered the serum hormones, morphologic and apoptotic measurements of genital organs in post-weaning gilts. *Asian-Australas J. Anim. Sci.* **2015**, *28*, 171–179. [CrossRef]

55. National Research Council. *Nutrient Requirements of Swine*; National Academies Press: Washington, DC, USA, 2012.

56. AOAC. *Official Methods of Analysis of the AOAC*; AOAC International: Rockville, MD, USA, 2012.

57. Jiang, S.Z.; Yang, Z.B.; Yang, W.R.; Wang, S.J.; Liu, F.X.; Johnston, L.A.; Chi, F.; Wang, Y. Effect of purified zearalenone with or without modified montmorillonite on nutrient availability, genital organs and serum hormones in post-weaning piglets. *Livest. Sci.* **2012**, *144*, 110–118. [CrossRef]

58. Rivera, A.; Agnati, L.F.; Horvath, T.L.; Valderrama, J.J.; Calle, A.D.L.; Fuxe, K. Uncoupling protein 2/3 immunoreactivity and the ascending dopaminergic and noradrenergic neuronal systems: Relevance for volume transmission. *Neuroscience* **2006**, *137*, 1447–1461. [CrossRef] [PubMed]

59. Zhou, M.; Yang, L.; Shao, M.; Wang, Y.; Yang, W.; Huang, L.; Zhou, X.; Jiang, S.; Yang, Z. Effects of zearalenone exposure on the TGF-beta1/Smad3 signaling pathway and the expression of proliferation or apoptosis related genes of post-weaning gilts. *Toxins* **2018**, *10*, 49. [CrossRef] [PubMed]

60. Livak, K.J.; Schmittgen, T.D. Analysis of relative gene expression data using real-time quantitative PCR and the 2 (-Delta Delta C(T)) method. *Methods* **2001**, *25*, 402–408. [CrossRef] [PubMed]

Article

Experimental Study on the Status of Maize Mycotoxin Production in Farmers' Grain Storage Silos in Northeastern China

Jinsong Zhang [1], Yan Xu [1], Taogang Hu [2], Changpo Sun [1,3,*] and Wenfu Wu [1,2,*]

[1] Department of Agricultural Engineering, College of Biological and Agricultural Engineering, Jilin University, Changchun 130022, China; jluzjs@jlu.edu.cn (J.Z.); xuyan@jlu.edu.cn (Y.X.)
[2] Department of Grain Science and Technology, Jilin Business and Technology College, Changchun 130507, China; lkyhtg@163.com
[3] Standards and Quality Center of National Food and Strategic Reserves Administration, No.11 Baiwanzhuang Avenue, Xicheng District, Beijing 100037, China
* Correspondence: chpsun@163.com (C.S.); wuwf@jlu.edu.cn (W.W.)

Abstract: The scientific rationality of farmers' grain storage technology and equipment is crucial for the biosecurity of grain in the main grain-producing areas represented by Northeast China. In this paper, four farmer grain storage mock silos of different widths were used as a means to track an experimental cycle of grain storage. The absolute water potential of corn in all four silos at the beginning of the experiment was greater than the absolute water potential of air, prompting moisture migration from the grain interior to the air and down to about 14%. Moisture was influenced by wind direction, and moisture decreased faster with better ventilation on both sides of the grain silos. Therefore, grain silo width has a significant effect on the drying effect under naturally ventilated conditions of maize ears. This research focused on the determination and assessment of mycotoxin contamination under farmers' storage grain conditions and analyzed the effect of silo structure on the distribution of mycotoxin contamination. When the width was too large, areas of high mycotoxin infection existed in the middle of the grain silo, and ventilation and tipping could be used to reduce the risk of toxin production. This study proved that reasonable farmer grain storage techniques and devices in Northeast China can effectively protect grain from mycotoxin contamination.

Keywords: farmers' grain storage silos; absolute water potential; ventilation and drying; mycotoxins; contamination distribution

Key Contribution: The scientific rationality of farmers' grain storage devices is significant in protecting grain from fungal toxins and promoting grain loss reduction. The central location of grain storage silos often has a high incidence of mycotoxins, and this risk should be reduced by rational design of the structure and dimensions of the storage silos.

Citation: Zhang, J.; Xu, Y.; Hu, T.; Sun, C.; Wu, W. Experimental Study on the Status of Maize Mycotoxin Production in Farmers' Grain Storage Silos in Northeastern China. *Toxins* **2021**, *13*, 741. https://doi.org/10.3390/toxins13110741

Received: 11 September 2021
Accepted: 12 October 2021
Published: 20 October 2021

1. Introduction

Maize is one of the most widely grown crops in the world and is cultivated in more than 170 countries and regions worldwide. As the world's most productive food crop, it can be used in a wide variety of food and industrial products and is also the predominant forage grain [1]. From 1967 to 2019, world maize production increased from 272 million tons to 1.11 billion tons. With the increase in world population, the demand for maize in developing countries will also increase significantly. As a basic staple food, maize of excellent quality needs to maintain high standards in terms of organoleptic, nutritional, and microbiological quality. However, nutrient and dry matter losses are usually caused by spoilage molds, and mycotoxin contamination may occur at the pre-harvest and post-harvest stages [2–4]. Important fungal toxins associated with maize include aflatoxins, which are produced by *Aspergillus flavus* (AFs); deoxynivalenol (DON), which belongs to the monoterpene group of toxins and is mainly produced by *Fusarium graminearum* and *Fusarium pink*, also known as vomitoxins due to their characteristic ability to induce vomiting in animals; fumonisins (FMs), produced by *Fusarium verticillioides* and *Fusarium proliferatum*; and zearalenone (ZEA), produced by *Fusarium graminearum* [5–7]. Climate and storage conditions have a significant impact on the production of mycotoxins, and if maize is grown in the tropics and subtropics with high temperatures and humidity, ear and grain rots caused by a variety of fungi are prevalent, causing farmers to suffer substantial economic losses [8]. If storage conditions are not well managed, the large increase in the number of insects and microorganisms in maize makes it more susceptible to fungal attack, thus greatly increasing the chance and extent of contamination by mycotoxins [2].

In response to the paradox of current and future food shortages, governments often seek to increase food availability by increasing food production. However, another important measure is to reduce food losses to balance the growing demand for food production, which often does not receive the attention it deserves [9–12]. In East and Southern Africa, for example, maize, the most important staple food, has experienced severe post-harvest losses in the past, resulting in reduced income for farmers [13]. Maize is infested with molds during storage, producing large amounts of fungal toxins, thereby losing its edible value [14]. Farmers still use many traditional storage methods to preserve grains, such as ground storage (ground-level grain), bags, baskets, or jars. However, these measures often do not guarantee the grain's protection from insects, pests, rodents, and molds [15].

Chinese farmers' grain storage accounts for about 50% of the total national grain production, about 250 million tons. The average grain storage per household is about 1200 kg, but there are large regional differences. In Jilin and Heilongjiang Provinces, which are the main grain-producing areas in the northeast, the average grain storage of farmers is more than 5000 kg, and in Liaoning Province, the average household storage is 3000–5000 kg. Due to the lack of suitable grain storage equipment and storage management techniques in most areas of the country, the level of storage is low, and a large amount of grain is lost due to factors such as rodents, insects, and mildew. According to a sample survey conducted by the National Grain Bureau of China, the average loss rate of grain stored by farmers nationwide is about 8%, with an annual loss of about 40 billion kg of grain. Among the main varieties of grain stored by farmers, corn has the highest loss rate, with an average of about 11%; rice is about 6.5% and wheat is about 4.7%. Therefore, safe storage is crucial, and it directly affects the overall quality and safety of grain as well as the income of the majority of farmers. The main causes of losses are mold (contributing about 30%) and insect damage (about 21%) [16]. Farmers' grain storage losses in the northeast are also more serious, with an average of about 10.2%. The poor storage conditions have caused a serious deterioration in grain quality, which poses a great potential hazard to food and food safety in China [17].

To facilitate grain storage, different grain storage devices have been designed for farmers around the world. The metal silo is a cylindrical structure made of galvanized iron. It has been shown to be effective in protecting harvested grain not only from storage insects but also from pests such as rodents, insects, and birds [18,19]. Although metal silos suffer from poor airtightness and high costs, they have become one of the key technologies for effective post-harvest management of grains, thereby improving food security for smallholder farmers [20]. In recent years, closed grain storage units have been increasingly promoted in Asia and Africa [21,22]. This device prevents moisture loss from the grain and limits gas exchange, thus changing the atmosphere inside the device. Purdue Improved Crop Storage (PICS) storage bags have been developed and promoted as a way to address the grain storage problems faced by farmers in developing countries. This is a sealing technology that works by strictly limiting the inflow of oxygen into the bulk grain. PICS bags can reduce the growth of insect populations in storage by 98% and can reduce grain losses due to insects and molds in storage to less than 1% while maintaining their quality for months or longer [23–25]. In addition to this, there are plastic silos [24], grain safety bags [26], and Grain Pro Super bags [27], which are containers with a multilayer composite technology for better gas-tight storage.

The most used grain storage devices in Northeast China are assembled corn ear storage silos. To meet the characteristics of rapid ventilation and water reduction, grain storage silos are designed as a single-side ventilated steel skeleton metal mesh structure. Such silos are conducive to natural ventilation and precipitation, and solve the problems of high moisture and easy mildew of newly harvested maize ears [28,29]. To save the land and increase corn storage capacity, corn ear storage silos can be designed with a hollow central silo structure by combining the "chimney effect" of air convection heat transfer with traditional natural ventilation [28]. The moisture content of corn at harvest is generally 25–33% and must be reduced to safe moisture before storage. The use of hot air drying can reduce the moisture in a short period, but corn treated with hot air has flaky kernels and reduced quality. Maize drying technology on silos has been greatly developed. Natural low-temperature grain storage can be achieved by taking natural ventilation and freezing precipitation in autumn and winter [30]. This can inhibit the respiration intensity of grain, delay aging, and reduce the loss of stored grain. Reasonable control of the in-silo and out-silo periods will affect the final quality of corn and the safety of grain storage.

Currently, there are few studies on the effects of farmer storage silos on maize mycotoxin production in Northeast China [31,32]. Therefore, this study conducted a natural ventilation test on corn ears to analyze water diffusion and temperature distribution of corn ears during natural precipitation and the effects of these factors on the production of mycotoxins. Based on the mining of biosecurity data from previous farmers' grain storage, it is shown that farmers' scientific grain storage can still adapt to the biosecurity requirements of grain in the new era and is an effective means to ensure the supply of high-quality grain materials to the market.

2. Results

2.1. Moisture Changes in Grain Piles

Table 1 shows the average moisture values of corn ears in each silo during the storage period. From the data in the table, it could be seen that the average moisture in silos 2, 3, and 4 had dropped to the safe moisture level specified in the national standard. Silo 1, which was wider and relatively poorly ventilated, also had its average moisture content reduced to 15.06% at the end of the test. Therefore, the use of rectangular silos with a reasonable structure for the storage of corn ears could reduce their moisture content to a safe moisture level after about 4 months of natural ventilation.

Table 1. Average moisture change during grain storage (unit: %).

Time	Silo 1	Silo 2	Silo 3	Silo 4
The second half of December	25.40	24.88	25.27	24.55
The first half of January	24.75	24.49	23.97	23.73
The second half of January	24.10	23.61	23.06	22.66
The first half of February	22.37	21.59	20.95	20.53
The second half of February	21.19	20.24	19.53	18.70
The first half of March	20.66	19.61	18.48	18.01
The second half of March	18.36	16.77	15.53	16.17
The first half of April	15.06	13.09	12.01	13.69

2.2. Variation and Distribution of Absolute Water Potential in Grain Piles

The concept of water potential was introduced into the field of grain drying in 2003 and applied to construct a grain drying model. It is mainly due to the fact that water migration is related to the existence of water potential difference between the inside and outside of grain particles. In 2007, Wenfu Wu numerically solved a water potential-based model for drying maize using a difference algorithm. The results of the solution were compared with experimental data and showed that the model could be applied to the simulation of the vacuum drying process of maize [33]. Later, the Laplace transform was further applied to derive an exact algebraic model from the dichotomous model of drying based on water potential [34], and a theoretical model of water potential applied to the moisture change process in naturally ventilated corn cob storage silos was established. In 2012, Zidan Wu et al. proposed a method to manage and control mechanical ventilation operations in grain silos using water potential maps. The concept and principle of water potential were further introduced into the management of the mechanical ventilation of grain silos by drawing an absolute water potential map of grain [35]. In 2016, Zhe Liu conducted a simulation and experimental study of the deep bed drying process of grain based on the conceptual model of water potential [36]. After the development of the above research work, the absolute water potential of air in the grain pile, E_{ja} (Equation (1)), and the absolute water potential of grain, E_{jg} (Equation (2)), were defined. For the correlation coefficients in the equations (i.e., A_1, A_2, B_1, B_2, and D_0), data from the equilibrium moisture isotherm model of maize were used [37,38]. These parameters were obtained by fitting desorption and adsorption equations by nonlinear regression analysis using self-determined equilibrium moisture/equilibrium relative humidity data of maize by the Academy of Sciences of China Grain Bureau. This provides a theoretical basis for grain drying, grain depot ventilation decision, and conditioning storage. The absolute water potential lines for air and grain can be plotted in the water potential diagram. The absolute water potentials of air and grain were used to calculate the absolute water potential of corn during natural ventilation and the absolute water potential of air under experimental conditions, respectively, to assess the absolute water potential variation and distribution of grain piles and the drying effect of grain silos.

The evaporation of liquid water vaporization inside the grain needs to gain energy and overcome the resistance of the grain for the moisture to migrate from the grain interior to reach the grain surface. Additionally, the evaporative migration of water from the surface of the grain to the air must overcome the binding energy between the water and the grain. In this study, based on the previous theoretical basis, the absolute water potential was used to characterize the water migration due to the energy exchange caused by any kind of unbalanced potential between the inside of the grain and the external environment, such as temperature, pressure, and moisture content. If the absolute water potential of the corn is greater than the absolute water potential of the ambient air, the grain moisture migrates to the air and the grain is in a state of desorption. If the absolute water potential of the grain is less than the absolute water potential of the air, the moisture in the air migrates into the grain kernels to produce adsorption.

The absolute water potential values of air and corn during the test were obtained using Equations (1) and (2), as shown in Tables 2 and 3. The trend of the absolute water potential of the corn and the external environment in each silo during the whole storage process is shown in Figure 1. At the beginning of the experiment, the absolute water potential of corn in all four silos was greater than the absolute water potential of air. Under the action of the water potential, the moisture migrated from the inside of the grain to the air, and the moisture content of the corn in the silos gradually decreased, with the average moisture content dropping from about 25% at the beginning to about 14%. As the ambient temperature rose, the absolute water potential of both air and corn gradually increased, but the difference between the absolute water potential of corn and the absolute water potential of air gradually decreased. When the absolute water potential of the corn was equal to the absolute water potential of the air, the water molecules did not have enough energy to diffuse from the surface to the surrounding air, and although the ambient temperature was higher the corn moisture was no longer falling.

Table 2. The absolute water potential energy of corn in each silo during natural ventilation (unit: kJ/kg).

Time	Silo 1	Silo 2	Silo 3	Silo 4
The second half of December	635.37	637.27	652.04	647.72
The first half of January	649.13	652.46	650.66	650.36
The second half of January	653.21	655.01	654.25	652.44
The first half of February	672.39	667.69	656.38	670.88
The second half of February	697.12	698.42	695.99	693.21
The first half of March	746.15	747.81	743.61	744.14
The second half of March	798.69	791.33	798.22	792.23
The first half of April	830.02	814.49	807.11	827.47

Table 3. The absolute water potential energy of air (unit: kJ/kg).

Time	Average Temperature (°C)	Average Humidity (%)	Average Wind Speed (m/s)	Average Absolute Potential Energy (kJ/kg)
The second half of December	−11.12	35	0.22	619.75
The first half of January	−11.21	33	0.32	615.21
The second half of January	−10.93	30	0.33	614.54
The first half of February	−9.50	28	0.78	626.48
The second half of February	−7.48	34	0.76	661.43
The first half of March	−3.11	37	0.43	716.78
The second half of March	1.49	46	1.14	783.38
The first half of April	6.38	33	1.19	822.14
Average value	−5.69	34	0.65	682.46

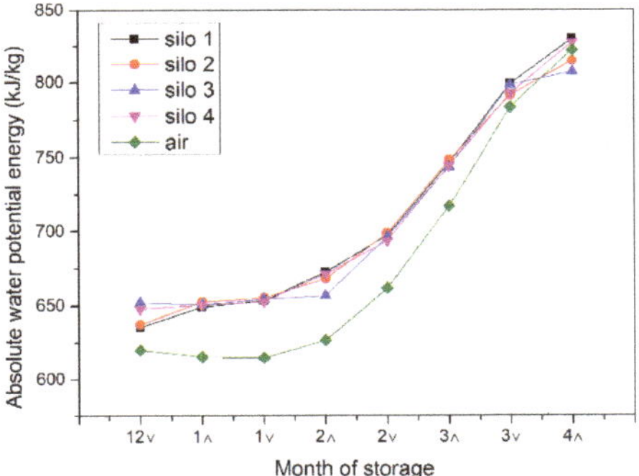

Figure 1. Comparison of water potential of corn in the silos with air water potential. The horizontal numbers represent months, "∧" represents the first half of the month, and "∨" represents the second half of the month.

2.3. Detection and Distribution of Mycotoxins in Grain Piles

In each of the four storage silos, 16 toxin sampling points were selected for each silo. Table 4 shows the results of aflatoxin B1 and B2 measurements on the collected samples (Before and after storage, aflatoxins G1 and G2 were not detected, so they were not shown in the table. Aflatoxin B1 and B2 were both non-detected before storage). Table 5 shows the results of the determination of deoxynivalenol (DON) content in the collected samples (413.24 ± 3.57 µg/kg of DON before storage). Table 6 shows the results of the determination of the zearalenone (ZEN) content of the collected samples (25.32 ± 1.36 µg/kg of ZEN before storage).

Table 4. Results of the determination of aflatoxin content in corn samples from storage silos (unit: μg/kg).

	Sample No.	AFB1	AFB2	Sample No.	AFB1	AFB2	Sample No.	AFB1	AFB2	Sample No.	AFB1	AFB2
Silo 1												
Top	1–1T	0.25 ± 0.01	/		/	/		/	/		/	/
Layer 1	1–1	0.35 ± 0.03	/	1–9	0.40 ± 0.04	0.05 ± 0.00	1–13	0.38 ± 0.02	/	1–21	0.32 ± 0.03	/
Layer 2	1–2	0.68 ± 0.07	0.10 ± 0.01	1–10	0.75 ± 0.02	0.43 ± 0.01	1–14	0.64 ± 0.03	0.45 ± 0.02	1–22	0.59 ± 0.01	/
Layer 3	1–3	0.86 ± 0.02	0.40 ± 0.04	1–11	0.95 ± 0.01	0.75 ± 0.03	1–15	0.90 ± 0.10	0.65 ± 0.05	1–23	0.85 ± 0.09	0.25 ± 0.01
Layer 4	1–4	0.72 ± 0.06	0.25 ± 0.02	1–12	0.78 ± 0.00	0.60 ± 0.04	1–16	0.68 ± 0.02	0.51 ± 0.02	1–24	0.62 ± 0.08	0.10 ± 0.00
Bottom	1–4B	0.48 ± 0.01	/	1–12B	0.62 ± 0.01	/	1–16B	0.58 ± 0.05	/	1–24B	0.45 ± 0.06	/
Silo 2												
Top	2–25T	/	/	/	/	/	/	/	/	/	/	/
Layer 1	2–25	0.30 ± 0.02	/	2–33	0.35 ± 0.01	/	2–37	0.32 ± 0.07	/	2–45	0.28 ± 0.00	/
Layer 2	2–26	0.58 ± 0.03	/	2–34	0.63 ± 0.04	0.05 ± 0.01	2–38	0.57 ± 0.05	0.10 ± 0.02	2–46	0.51 ± 0.08	/
Layer 3	2–27	0.71 ± 0.07	0.11 ± 0.01	2–35	0.82 ± 0.12	0.49 ± 0.10	2–39	0.75 ± 0.10	0.55 ± 0.08	2–47	0.68 ± 0.04	0.05 ± 0.01
Layer 4	2–28	0.62 ± 0.05	/	2–36	0.67 ± 0.10	0.35 ± 0.06	2–40	0.65 ± 0.08	0.46 ± 0.10	2–48	0.60 ± 0.07	/
Bottom	2–28B	0.48 ± 0.03	/	2–36B	0.48 ± 0.06	/	2–40B	0.46 ± 0.06	/	2–48B	0.39 ± 0.01	/
Silo 3												
Top	3–49T	/	/		/	/		/	/		/	/
Layer 1	3–49	/	/	3–57	/	/	3–61	/	/	3–65	/	/
Layer 2	3–50	0.28 ± 0.01	/	3–58	0.55 ± 0.03	0.04 ± 0.00	3–62	0.49 ± 0.03	/	3–66	0.45 ± 0.04	/
Layer 3	3–51	0.40 ± 0.03	0.11 ± 0.01	3–59	0.70 ± 0.06	0.35 ± 0.07	3–63	0.65 ± 0.11	0.25 ± 0.06	3–67	0.60 ± 0.09	0.05 ± 0.01
Layer 4	3–52	0.32 ± 0.01	/	3–60	0.60 ± 0.04	0.10 ± 0.01	3–64	0.58 ± 0.09	0.03 ± 0.01	3–68	0.55 ± 0.04	/
Bottom	3–52B	/	/	3–60B	0.26 ± 0.01	/	3–64B	0.24 ± 0.02	/	3–68B	/	/
Silo 4												
Top	4–69T	/	/		/	/		/	/		/	/
Layer 1	4–69	/	/	4–73	/	/	4–77	/	/	4–81	/	/
Layer 2	4–70	0.20 ± 0.01	/	4–74	0.44 ± 0.03	/	4–78	0.36 ± 0.02	/	4–82	0.28 ± 0.02	/
Layer 3	4–71	0.35 ± 0.03	/	4–75	0.50 ± 0.08	0.17 ± 0.01	4–79	0.45 ± 0.07	/	4–83	0.30 ± 0.02	/
Layer 4	4–72	0.24 ± 0.00	/	4–76	0.34 ± 0.02	0.10 ± 0.00	4–80	0.28 ± 0.01	/	4–84	0.30 ± 0.03	/
Bottom	4–72B	/	/	4–76B	/	/	4–80B	/	/	4–84B	/	/

Note: "/" in the table means not detected. AFG1 and AFG2 were not detected before and after the experiment, so they are not reflected in the table.

Table 5. Results of the determination of deoxynivalenol (DON) content in corn samples from grain storage silos (unit: µg/kg).

	Sample No.	DON	Sample No.	DON	Sample No.	DON	Sample No.	DON
Silo 1								
Top	1–1T	707.92 ± 0.81						
Layer 1	1–1	714.20 ± 3.95	1–9	718.96 ± 5.01	1–13	721.46 ± 5.83	1–21	710.68 ± 4.49
Layer 2	1–2	715.46 ± 7.01	1–10	724.50 ± 8.91	1–14	733.62 ± 8.63	1–22	719.80 ± 6.61
Layer 3	1–3	721.31 ± 8.56	1–11	730.81 ± 9.14	1–15	741.51 ± 7.91	1–23	727.65 ± 8.19
Layer 4	1–4	718.78 ± 4.89	1–12	726.43 ± 6.01	1–16	731.63 ± 5.68	1–24	720.77 ± 6.74
Bottom	1–4B	694.34 ± 6.90	1–12B	689.85 ± 6.81	1–16B	701.49 ± 7.77	1–24B	705.38 ± 6.36
Silo 2								
Top	2–25T	689.30 ± 0.61						
Layer 1	2–25	701.68 ± 7.91	2–33	696.53 ± 6.83	2–37	690.64 ± 7.66	2–45	684.11 ± 7.80
Layer 2	2–26	708.76 ± 7.89	2–34	713.54 ± 8.43	2–38	700.35 ± 8.02	2–46	695.40 ± 8.26
Layer 3	2–27	715.41 ± 9.25	2–35	736.26 ± 11.05	2–39	720.50 ± 11.89	2–47	711.47 ± 11.29
Layer 4	2–28	712.53 ± 10.27	2–36	722.87 ± 9.31	2–40	716.96 ± 8.96	2–48	700.84 ± 8.70
Bottom	2–28B	684.26 ± 7.13	2–36B	675.11 ± 8.31	2–40B	670.24 ± 7.73	2–48B	666.21 ± 7.87
Silo 3								
Top	3–49T	645.42 ± 1.71						
Layer 1	3–49	650.33 ± 6.11	3–57	659.22 ± 6.52	3–61	664.16 ± 5.93	3–65	653.20 ± 6.12
Layer 2	3–50	657.15 ± 6.00	3–58	665.31 ± 6.85	3–62	670.20 ± 6.06	3–66	658.67 ± 5.45
Layer 3	3–51	675.30 ± 8.81	3–59	685.45 ± 9.06	3–63	689.75 ± 7.88	3–67	670.76 ± 9.37
Layer 4	3–52	666.47 ± 8.39	3–60	672.91 ± 8.15	3–64	680.03 ± 9.05	3–68	660.36 ± 8.29
Bottom	3–52B	620.94 ± 5.96	3–60B	629.47 ± 6.69	3–64B	630.15 ± 7.75	3–68B	616.23 ± 6.60
Silo 4								
Top	4–69T	620.44 ± 0.77						
Layer 1	4–69	653.02 ± 9.97	4–73	661.33 ± 12.16	4–77	641.42 ± 9.33	4–81	638.37 ± 10.54
Layer 2	4–70	655.41 ± 12.03	4–74	669.23 ± 11.85	4–78	649.75 ± 12.61	4–82	640.14 ± 12.19
Layer 3	4–71	660.44 ± 13.17	4–75	680.35 ± 11.99	4–79	670.99 ± 13.93	4–83	650.15 ± 13.15
Layer 4	4–72	658.86 ± 12.77	4–76	674.36 ± 13.72	4–80	650.65 ± 13.28	4–84	642.82 ± 14.11
Bottom	4–72B	618.29 ± 7.81	4–76B	611.82 ± 6.58	4–80B	607.10 ± 7.99	4–84B	600.33 ± 7.62

Table 6. Results of the determination of zearalenone (ZEN) content in corn samples from grain storage silos (unit: μg/kg).

	Sample No.	ZEN	Sample No.	ZEN	Sample No.	ZEN	Sample No.	ZEN
Silo 1								
Top	1–1T	36.61 ± 1.24						
Layer 1	1–1	37.32 ± 0.08	1–9	37.35 ± 0.26	1–13	36.93 ± 0.22	1–21	37.22 ± 0.20
Layer 2	1–2	39.34 ± 2.13	1–10	43.45 ± 1.99	1–14	42.45 ± 1.37	1–22	41.21 ± 1.75
Layer 3	1–3	40.12 ± 3.01	1–11	47.09 ± 4.25	1–15	46.11 ± 3.93	1–23	40.25 ± 3.65
Layer 4	1–4	38.28 ± 0.41	1–12	39.00 ± 0.72	1–16	38.75 ± 0.43	1–24	37.78 ± 0.72
Bottom	1–4B	36.30 ± 0.66	1–12B	37.46 ± 1.07	1–16B	36.67 ± 0.56	1–24B	35.48 ± 0.91
Silo 2								
Top	2–25T	34.41 ± 2.03						
Layer 1	2–25	34.32 ± 0.63	2–33	35.21 ± 0.46	2–37	34.89 ± 0.49	2–45	34.08 ± 0.46
Layer 2	2–26	35.27 ± 0.74	2–34	36.38 ± 0.57	2–38	35.88 ± 0.65	2–46	34.98 ± 0.56
Layer 3	2–27	37.94 ± 3.06	2–35	42.11 ± 2.69	2–39	38.67 ± 2.08	2–47	36.45 ± 1.69
Layer 4	2–28	35.40 ± 1.81	2–36	38.24 ± 0.94	2–40	36.42 ± 1.26	2–48	35.11 ± 1.59
Bottom	2–28B	33.09 ± 0.01	2–36B	33.04 ± 0.02	2–40B	32.67 ± 0.01	2–48B	32.23 ± 0.03
Silo 3								
Top	3–49T	33.34 ± 0.98						
Layer 1	3–49	34.67 ± 0.44	3–57	34.22 ± 0.31	3–61	34.55 ± 0.39	3–65	33.80 ± 0.42
Layer 2	3–50	35.10 ± 1.03	3–58	36.67 ± 0.88	3–62	36.17 ± 0.93	3–66	34.63 ± 1.04
Layer 3	3–51	35.15 ± 2.77	3–59	40.03 ± 3.07	3–63	38.14 ± 1.53	3–67	35.34 ± 2.07
Layer 4	3–52	34.86 ± 2.37	3–60	38.81 ± 2.49	3–64	36.70 ± 1.83	3–68	34.22 ± 1.63
Bottom	3–52B	33.10 ± 0.41	3–60B	33.45 ± 0.19	3–64B	33.44 ± 0.32	3–68B	32.91 ± 0.20
Silo 4								
Top	4–69T	31.56 ± 1.01						
Layer 1	4–69	31.06 ± 0.51	4–73	31.78 ± 0.39	4–77	32.01 ± 0.49	4–81	31.21 ± 0.53
Layer 2	4–70	31.91 ± 1.01	4–74	33.85 ± 2.03	4–78	32.93 ± 0.68	4–82	31.50 ± 0.40
Layer 3	4–71	32.72 ± 1.89	4–75	35.81 ± 2.07	4–79	36.05 ± 1.33	4–83	32.22 ± 2.71
Layer 4	4–72	32.15 ± 1.65	4–76	34.62 ± 1.09	4–80	34.38 ± 2.05	4–84	31.83 ± 1.13
Bottom	4–72B	30.45 ± 0.72	4–76B	31.74 ± 0.57	4–80B	30.78 ± 0.62	4–84B	30.09 ± 0.81

2.3.1. Aflatoxin Contents and Distribution

From Table 4, it could be calculated that the total aflatoxin content of each testing point in each storage silo was between 0 and 1.70 µg/kg. The safe limit range of total aflatoxin content in food in China is <5–10 µg/kg, so the aflatoxin is within the safe level, and the total aflatoxin contamination level can be effectively controlled in farmers' grain storage silos. Table 7 reflects the average aflatoxin content of each layer in each storage silo. The data for each layer were summed and averaged to obtain the average aflatoxin content. From Table 8, it could be seen that silo 4 among the four storage silos had the best effect in controlling the contamination level of total aflatoxin content. Combined with Figure 2, it also could be seen that the middle layer (i.e., the third layer) and the middle column samples of all silos had higher aflatoxin contents. There were significant differences in aflatoxin levels between the layers of the silos, with a difference of 75.14% between the minimum and maximum values (Table 9). The average aflatoxin level in the third layer of silo 1 reached a maximum of 1.403 µg/kg; therefore, the middle area was more likely to contribute to aflatoxin production during storage.

Table 7. Average aflatoxin levels in each layer of grain storage silos (unit: µg/kg).

	Silo 1	Silo 2	Silo 3	Silo 4
Top	0.06 ± 0.01	0	0	0
Layer 1	0.38 ± 0.03	0.31 ± 0.01	0	0
Layer 2	0.88 ± 0.07	0.61 ± 0.05	0.45 ± 0.03	0.32 ± 0.02
Layer 3	1.40 ± 0.09	1.04 ± 0.11	0.78 ± 0.11	0.44 ± 0.05
Layer 4	1.07 ± 0.07	0.85 ± 0.09	0.55 ± 0.05	0.32 ± 0.02
Bottom	0.53 ± 0.03	0.45 ± 0.03	0.13 ± 0.01	0

Table 8. Levels of aflatoxin in each column of the grain storage silos (unit: µg/kg).

	Silo 1	Silo 2	Silo 3	Silo 4
West 1 column	0.45 ± 0.03	0.31 ± 0.02	0.12 ± 0.01	0.09 ± 0.01
West 2 column	0.59 ± 0.02	0.43 ± 0.06	0.29 ± 0.03	0.17 ± 0.02
West 3 column	0.60 ± 0.04	0.48 ± 0.07	0.28 ± 0.04	0.14 ± 0.01
West 4 column	0.45 ± 0.04	0.36 ± 0.03	0.24 ± 0.03	0.13 ± 0.01

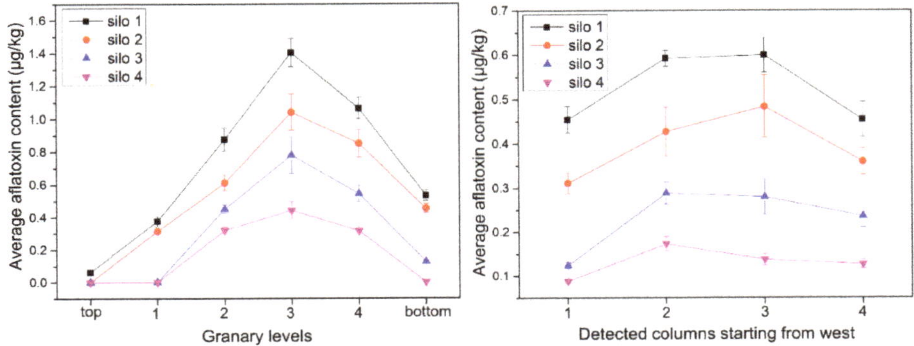

Figure 2. Aflatoxin contents of samples from different locations of grain storage silos.

Table 9. The average contents of aflatoxin in each storage silo and the ratio of difference with the content in silo 1.

Indicators	Silo 1	Silo 2	Silo 3	Silo 4
Toxin content (µg/kg)	0.36 ± 0.02a	0.27 ± 0.03b	0.16 ± 0.02c	0.09 ± 0.01d
Ratio of difference (%)	0	25.14	56.08	75.14

Note: The differences between the values with different letters in the same row are significant ($p < 0.05$).

2.3.2. Deoxynivalenol (DON) Contents and Distribution

As can be seen from Table 5, the DON content of each testing point of each grain storage silo was between 600 and 750 µg/kg. The safe limit range of DON in food is <1000 µg/kg in China, so the DON content of each silo is within the safe level, and the farmers' grain storage silo can effectively control DON contamination. Table 10 reflects the average DON content of each layer in each storage silo. The results obtained for each layer in Table 10 were summed and averaged to obtain the average DON contents in Table 11. The data showed that silo 4 had the best effect in controlling the DON contamination level. As can be seen in Figure 3, the DON content of the middle (third) sample was higher. Although silo 1 had the highest level of DON content, there was no significant difference in DON contents between the silos, and the difference between the minimum and maximum values was only 10.05% of the ratio (Table 12). This indicated that DON was mainly produced before entering the silos (DON content was 413.24 ± 3.57 µg/kg before storage), and it was advisable to use measures such as timely harvesting and drying into the silos to mitigate mycotoxin production.

Table 10. Average deoxynivalenol levels in each layer of grain storage silos (unit: µg/kg).

	Silo 1	Silo 2	Silo 3	Silo 4
Top	707.92 ± 0.81	689.30 ± 0.61	645.42 ± 1.71	620.44 ± 0.77
Layer 1	716.33 ± 4.82	693.24 ± 7.55	656.73 ± 6.17	648.54 ± 10.50
Layer 2	723.35 ± 7.79	704.51 ± 8.15	662.83 ± 6.09	653.63 ± 12.17
Layer 3	730.32 ± 8.45	720.91 ± 10.87	680.32 ± 8.78	665.48 ± 13.06
Layer 4	724.40 ± 5.83	713.30 ± 9.31	669.94 ± 8.47	656.67 ± 13.47
Bottom	697.77 ± 6.96	673.96 ± 7.76	624.20 ± 6.75	609.39 ± 7.50

Table 11. Contents of deoxynivalenol in each column of the grain storage silos (unit: µg/kg).

	Silo 1	Silo 2	Silo 3	Silo 4
West 1 column	712.00 ± 5.35	700.82 ± 7.18	652.60 ± 6.16	644.41 ± 9.51
West 2 column	718.11 ± 7.18	708.86 ± 8.79	663.98 ± 5.96	659.42 ± 11.26
West 3 column	725.94 ± 7.16	699.74 ± 8.85	667.36 ± 6.84	643.98 ± 11.43
West 4 column	716.86 ± 6.48	691.61 ± 8.79	651.84 ± 7.16	634.46 ± 11.42

Table 12. The average contents of deoxynivalenol in each storage silo and the ratio of difference with the content in silo 1.

Indicators	Silo 1	Silo 2	Silo 3	Silo 4
Toxin content (µg/kg)	717.91 ± 12.95a	700.63 ± 18.31a	658.10 ± 20.25b	645.42 ± 22.68b
Ratio of difference (%)	0	2.12	8.23	10.05

Note: The differences between the values with different letters in the same row are significant ($p < 0.05$).

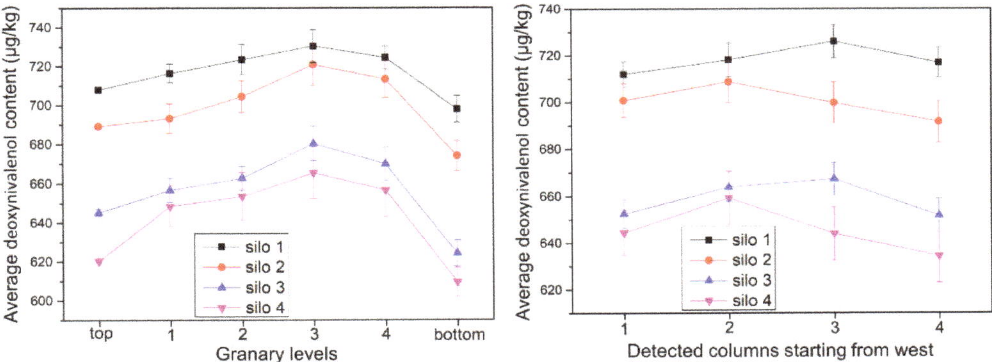

Figure 3. Deoxynivalenol contents of samples from different locations of grain storage silos.

2.3.3. Zearalenone (ZEN) Contents and Distribution

From Table 6, it could be seen that the ZEN content of each testing point was between 30 and 47 µg/kg. The safe limit range of ZEN in food is <60 µg/kg in China, so the ZEN content of each silo is within the safe level. Table 13 reflects the average content of ZEN in each layer in each storage silo. The data for each layer were summed and averaged to obtain the average level of ZEN content, as shown in Table 14, which showed that silo 4 has the best effect in controlling the ZEN contamination level. As can be seen in Figure 4, the ZEN content of the sample from the middle layer (third layer) was higher. There is a significant difference between the ZEN content levels in each layer of the silo, with a minimum and maximum difference ratio of 17% (Table 15).

Table 13. Average zearalenone contents in each layer of grain storage silos (unit: µg/kg).

	Silo 1	Silo 2	Silo 3	Silo 4
Top	36.61 ± 1.24	34.41 ± 2.03	33.34 ± 0.98	31.56 ± 1.01
Layer 1	37.21 ± 0.19	34.63 ± 0.51	34.31 ± 0.39	31.52 ± 0.48
Layer 2	41.61 ± 1.81	35.63 ± 0.63	35.64 ± 0.97	32.55 ± 1.03
Layer 3	43.39 ± 3.71	38.79 ± 2.38	37.17 ± 2.36	34.20 ± 2.00
Layer 4	38.45 ± 0.57	36.29 ± 1.40	36.15 ± 2.08	33.25 ± 1.48
Bottom	36.48 ± 0.80	32.76 ± 0.36	33.23 ± 0.28	30.77 ± 0.68

Table 14. Contents of zearalenone in each column of the grain storage silos (unit: µg/kg).

	Silo 1	Silo 2	Silo 3	Silo 4
West 1 column	38.00 ± 1.25	35.07 ± 1.38	34.37 ± 1.33	31.64 ± 1.13
West 2 column	40.87 ± 1.66	37.00 ± 0.94	36.64 ± 1.39	33.56 ± 1.23
West 3 column	40.18 ± 1.30	35.71 ± 0.90	35.80 ± 1.00	33.23 ± 1.03
West 4 column	38.39 ± 1.45	34.60 ± 0.90	34.18 ± 1.07	31.37 ± 1.12

Table 15. The average contents of zearalenone in each storage silo and the ratio of difference with the content in silo 1.

Indicators	Silo 1	Silo 2	Silo 3	Silo 4
Toxin content (µg/kg)	38.96 ± 3.18a	35.42 ± 2.29ab	34.97 ± 1.93ab	32.31 ± 1.64b
Ratio of difference (%)	0	9.62	10.48	17.68

Note: The differences between the values with different letters in the same row are significant ($p < 0.05$).

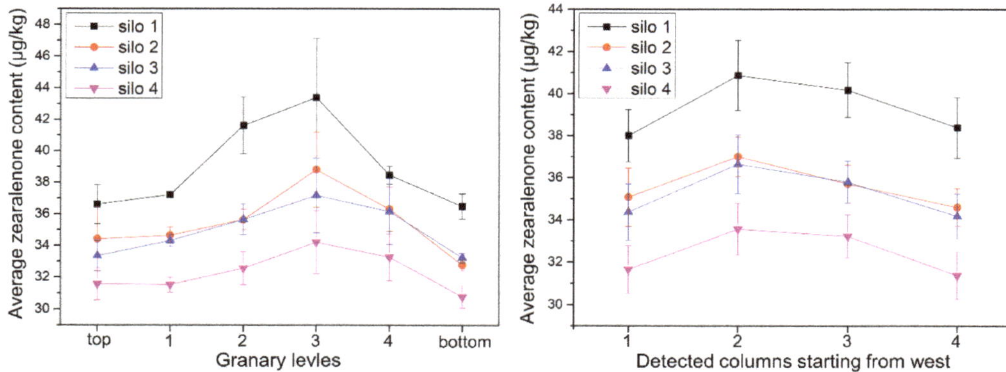

Figure 4. Zearalenone contents of samples from different locations of grain storage silos.

2.4. Mold Rate

The Chinese national standard GB1353–2009 makes clear provisions for mold rate, and raw mold grains need to be ≤2%. Table 16 shows the mold rate of corn in the silos at the end of the experiment. The data indicated that the mold rate in silos 1 and 2 was relatively high and not uniform, while both indicators were better in silos 3 and 4. It could be seen that the structure of the silo had a greater impact on the mold rate. The wider the width of the silo, the more serious the mold, and the fluctuation range of the mold rate was large. Therefore, the structure of the grain storage silo should be reasonably designed to reduce the occurrence of mold. The level of corn mycotoxin content was mainly related to whether the corn was infected with toxin-producing molds. Although there was no significant positive correlation between the number of moldy grains and mycotoxin content, the mold rate could be used to evaluate whether the storage silo design was reasonable from the perspective of whether maize was susceptible to mold infection [39,40].

Table 16. The rate of corn mold in the silo at the time of leaving the warehouse (unit: %).

	Silo 1	Silo 2	Silo 3	Silo 4
Maximum value	0.387	0.547	0.143	0.248
Minimum value	0	0.009	0	0
Difference	0.387	0.539	0.143	0.248
Average value	0.088	0.125	0.053	0.062

3. Discussion

The vaporization and evaporation of liquid water inside the grain require the acquisition of energy to overcome the internal resistance of the grain. Only then can the moisture migrate from inside the grain to reach the surface of the grain. Additionally, the evaporative migration of moisture from the grain surface to the air must overcome the binding energy between the moisture and the grain. The presence of an imbalance between the inside of the maize kernel and the external environment in terms of temperature, pressure, and moisture can cause energy exchange and result in water migration. In this study, absolute water potential was used to characterize this energy exchange. The activity of free water in corn ears in a grain silo cannot be measured directly. The strategy of this study was to calculate the average water change throughout the storage silo by detecting the initial water at the time the grain enters the silo, combined with real-time monitoring of the overall weight drop of the silo. The water potential was calculated by measuring the moisture distribution of each layer at the exit of the silo, and the water potential was positively correlated with the free water activity. If the absolute water potential of the corn was greater than the absolute water potential of the ambient air, the grain moisture migrated to the air and the grain was in a state of desorption. If the situation was reversed, the moisture in the air migrated into the grain kernels to produce adsorption.

At the end of the test, the gradient of water potential in each silo was obvious. Moisture was affected by wind direction, with better ventilation on both sides and a faster rate of moisture decline [41]. Tables 17–20 show the accumulated temperature values collected by 84 temperature sensors in the four grain storage silos. These temperature sensors were distributed in different locations of the grain silos (as shown in Figure 5). By examining the accumulated temperatures, it can be broadly observed that the accumulated temperatures are relatively high in the center of the silos. The corn in the middle had a higher moisture content, which, combined with the fact that the temperature in the middle was also high, made it easy to lead to the production of fungal toxins. It could be seen that the width of the grain silo (grain layer thickness) had an impact on the water potential distribution of corn under natural ventilation conditions, and an unreasonable silo structure could lead to high moisture concentration zones. The high moisture and high heat zone could be predicted by monitoring the absolute water potential in the silo during the grain storage period to ensure the safety of grain storage.

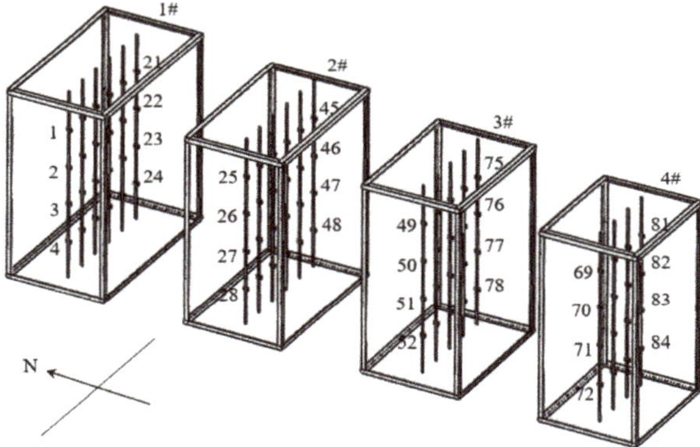

Figure 5. Sensor distribution in grain storage silos (84 temperature sensors labeled 1–84 in four silos).

Table 17. Values of the accumulated temperature obtained by temperature sensors at different points in grain storage silo 1 (unit: °C).

Sample No.	Accumulated Temperature	Sample No.	Accumulated Temperature	Sample No.	Accumulated Temperature	Sample No.	Accumulated Temperature	Sample No.	Accumulated Temperature	Sample No.	Accumulated Temperature	Sample No.	Accumulated Temperature
1	−2013.1	5	−2012.4	9	−2000.3	13	−2064.0	17	−2164.6	21	−2142.7		
2	−2137.6	6	−2120.6	10	−2084.8	14	−2104.3	18	−2253.9	22	−2279.4		
3	−2271.6	7	−2217.7	11	−2181.7	15	−2166.0	19	−2157.4	23	−2203.9		
4	−2328.7	8	−2310.5	12	−2225.6	16	−2258.9	20	−2283.6	24	−2268.7		

Table 18. Values of the accumulated temperature obtained by temperature sensors at different points in grain storage silo 2 (unit: °C).

Sample No.	Accumulated Temperature	Sample No.	Accumulated Temperature	Sample No.	Accumulated Temperature	Sample No.	Accumulated Temperature	Sample No.	Accumulated Temperature	Sample No.	Accumulated Temperature
25	−2026.5	29	−2035.6	33	−2028.0	37	−2034.9	41	−2084.3	45	−2149.3
26	−2127.9	30	−2119.7	34	−2117.9	38	−2073.3	42	−2115.1	46	−2106.7
27	−2216.3	31	−2194.1	35	−2137.6	39	−2166.4	43	−2112.7	47	−2180.3
28	−2252.9	32	−2232.4	36	−2244.1	40	−2209.1	44	−2179.2	48	−2197.7

Table 19. Values of the accumulated temperature obtained by temperature sensors at different points in grain storage silo 3 (unit: °C).

Sample No.	Accumulated Temperature	Sample No.	Accumulated Temperature	Sample No.	Accumulated Temperature	Sample No.	Accumulated Temperature	Sample No.	Accumulated Temperature
49	−2139.4	53	−2018.4	57	−1981.6	61	−2019.6	65	−2052.8
50	−2139.1	54	−2111.2	58	−2110.1	62	−2091.2	66	−2061.8
51	−2169.0	55	−2174.4	59	−2152.3	63	−2134.7	67	−2217.1
52	−2263.8	56	−2244.4	60	−2173.0	64	−2160.6	68	−2043.5

Table 20. Values of the accumulated temperature obtained by temperature sensors at different points in grain storage silo 4 (unit: °C).

Sample No.	Accumulated Temperature	Sample No.	Accumulated temperature	Sample No.	Accumulated Temperature	Sample No.	Accumulated Temperature
69	−2045.7	73	−2045.2	77	−2036.2	81	−2065.8
70	−2185.1	74	−2159.9	78	−2146.4	82	−2045.1
71	−2207.1	75	−2178.1	79	−2134.7	83	−2096.7
72	−2249.6	76	−2205.5	80	−2201.1	84	−2135.5

The main contributing factors to the increase in corn mycotoxins in farmers' grain storage silos include temperature, wind speed, corn ear precipitation rate, and dryness. If the temperature and humidity of agro products are too high, an increase in toxins is likely to occur. Under natural ventilation, the precipitation rate of the corn ear is mainly influenced by temperature and wind speed. During the grain storage process, the temperature changes of the grain piles in the four silos were not very different. During the natural ventilation process, the trend of the precipitation rate curve of grain was increasing with the ambient wind speed, temperature, and absolute water potential curve of grain [41]. Among them, the wind speed had the greatest effect on the precipitation rate of the grain piles. The trends of ambient wind speed curve and precipitation rate curve were generally consistent, both showing low values in January and February. Since the interior of the silo was a large hysteresis system, the overall precipitation rate curve of the grain inside the silo lagged behind the ambient wind speed curve. The absolute water potential of both grain and air increased as the temperature inside the silo gradually increased from $-15\ ^\circ\mathrm{C}$ to $5\ ^\circ\mathrm{C}$. The trend of the absolute water potential of corn was the same as that of the ambient temperature, indicating that temperature had a large influence on the absolute water potential. The difference in width caused a large difference in the precipitation rate curves of the four silos. In general, the smaller the width, the greater the precipitation rate. This is more conducive to the drying of the corn ear, thus inhibiting the increase in mycotoxins. Regarding the mycotoxin contamination of maize in Jilin Province that year, aflatoxin B1 was not detected; the highest value of deoxynivalenol was 966.9 µg/kg, with a mean value of 108.2 µg/kg; and the highest value of zearalenone was 42.2 µg/kg, with a mean value of 0.004 µg/kg (according to the Jilin Harvest Grain Quality Survey report). The toxin levels in this experiment were all less than the highest values reported. Mycotoxins in grain silos showed the characteristics of "inverted U-shaped" distribution, indicating that aflatoxins and zearalenone had a tendency to increase in storage and were sensitive to the width of grain silos. For example, regarding the concentration of ZEN in the distribution area of silo 1, the toxin level was close to the maximum allowable value, indicating that there was still a risk of grain storage in farmers' grain storage silos. Grain storage monitoring and control should be strengthened, and the width of the grain silo design should not be too large (preferably not more than 1.5 m in this study).

4. Conclusions

When using rectangular steel mesh ventilated grain storage silos for corn ear storage, the moisture content of corn ears could be reduced to a safe moisture level after 4 months of natural ventilation. This shows that farmers' scientific grain storage silos can ensure the safety requirements of stored grain mycotoxins under reasonable structure and normal year conditions. The absolute water potential of corn in all four silos at the beginning of the experiment was greater than the absolute water potential of air. This prompted the migration of moisture from the interior of the grain to the air, and the moisture of the corn in the silos gradually decreased to about 14%. The ambient temperature rises so that the absolute water potential of both air and corn gradually increases, but the difference between the two gradually decreases, and finally reaches an equal. Water molecules do not have enough energy to diffuse from the surface into the surrounding air. Moisture is affected by the wind direction, so with better ventilation on both sides, the moisture falls faster. Therefore, the width of the grain silo affects the distribution of water potential and the drying effect of corn under natural ventilation conditions, which in turn affects the production of mycotoxins. For the detection of three mycotoxins in the grain silo, it was shown that the production of mycotoxins was related to the structure of the silos. When the width is too large, there are areas of concentrated infection of mycotoxins such as AFT, DON, and ZEN. Therefore, a reasonable range of silo dimensions should be fully considered in the design of the parameters of grain storage silos. In the middle of the grain silo, there is a potential risk area. It is appropriate to take mechanical ventilation, mechanical turning, and other measures to destroy the opportunity and degree of toxin production in advance. For the sake of food security, scientific grain storage technology and techniques for farmers should be vigorously developed, and this policy fosters the application of scientific grain storage technology for farmers. The above conclusions apply only to the mid-temperate zone at 40–55 degrees north latitude.

5. Materials and Methods

5.1. Materials

The maize variety was Centaur 335, the largest planted in Jilin Province, with an initial moisture of 26%. The collected maize cobs were mixed samples from the edge and middle of the farm field. They were peeled and left naturally for 48 h to remove ears with lesions, mold, and damage. To ensure that all corn ears in the storage silos had the same level of mycotoxins, the ears were well mixed before entering the silos. Three points in the mixed grain pile were selected for sampling. The content of the three mycotoxins (Aflatoxin, Deoxynivalenol, and Zearalenone) and the rate of mildew were measured separately, and the average value was taken as the content of maize mycotoxins before storage. All samples were stored in silos.

Main reagents and consumables: methanol (analytical purity), acetonitrile (analytical purity), sodium chloride (chemical purity), polyethylene glycol 8000 (analytical purity), glass fiber filter paper, aflatoxin immunoaffinity column (Clover Technology Group Inc., Beijing, China), deoxynivalenol immunoaffinity column (Clover Technology Group Inc., Beijing, China), zearalenone immunoaffinity column (Clover Technology Group Inc., Beijing, China), aflatoxin mixed standard (Sigma-Aldrich China Ltd., Shanghai, China), deoxynivalenol standard (Sigma-Aldrich China Ltd., Shanghai, China), zearalenone standard (Sigma-Aldrich China Ltd., Shanghai, China), phosphate buffer for column crossing, 1.5 mL liquid phase vial set (Agilent), 0.45 um organic phase filtration membrane, 1 mL disposable syringe.

Main instruments and equipment: High performance liquid chromatograph Agilent 1260 with fluorescence and UV detector, pulverizer, high-speed homogenizer, nitrogen blowing apparatus, air pressure pump and pump flow rack, analytical balance (sensitivity 0.001 g), 20 mL glass syringe, 1 mm pore size test sieve.

5.2. Homemade Test Silos and Testing Systems

Four homemade naturally ventilated rectangular corn grain storage silos were used for the experiment (Figure 6). The distribution of many sensors inside them is shown in Figure 5. For the four rectangular steel mesh ventilation silos with different widths, length × width × height were 1 m × 1.8 m × 2 m, 1 m × 1.6 m × 2 m, 1 m × 1.4 m × 2 m, and 1 m × 1.2 m × 2 m, respectively. The tare weights of silos No. 1 to No. 4 were 115.44 kg, 106.08 kg, 96.72 kg, and 87.36 kg, respectively, loaded with 1.85 t, 1.56 t, 1.40 t, and 1.26 t of corn cobs. To test the grain storage effect of various silo sizes in actual operation, the shape of the silo with left and right closed impermeable panels and front and rear permeable mesh panels was used, making the width of the ventilated silo the main factor affecting the grain storage quality (width for left and right closed impermeable panels and length for those with mesh panels).

The testing system mainly included: weighing sensor and meter (Model TQ-ST02, Beijing Shitong Sci-Tech Co., Ltd., Beijing, China), grain moisture measuring instrument (Model PM-8188, KETT, Japan), temperature patrol meter (TR-4, homemade), temperature sensor (DS18B20).

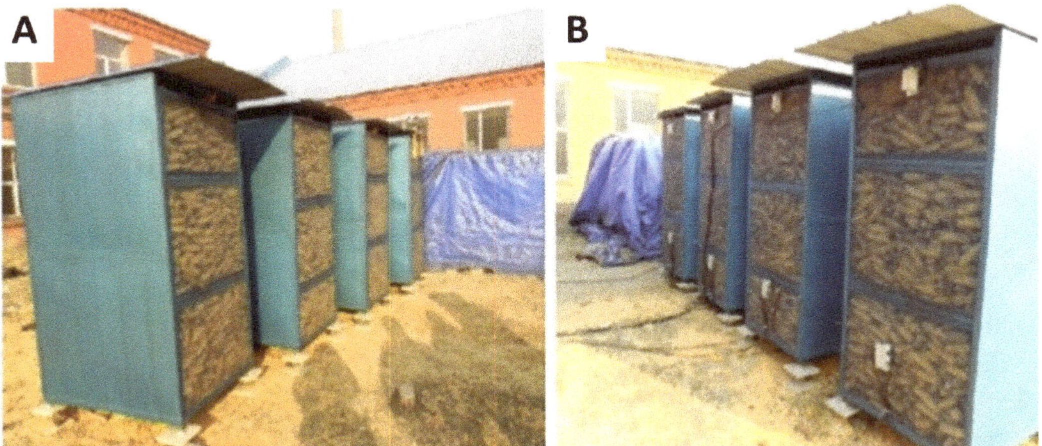

Figure 6. Natural ventilation corn ear storage silos. (**A**) Rectungular steel mesh facing east; (**B**) Rectungular steel mesh facing west.

5.3. Test Method

The test silos were placed outdoors in a north–south direction. Load cells were added to the foot of the silo to record the overall weight of the silo in real time. The moisture content of the corn in the silo was calculated indirectly through the change in the silo weight. There were 84 temperature sensors distributed in the four grain storage silos, and the numbers were: 24 in silo 1 (the widest), 24 in silo 2, 20 in silo 3, and 16 in silo 4 (the narrowest). The order of arrangement and location was from top to bottom, from west to east. The test started on December 16 of the previous year and ended on April 15 of the following year, with a storage period of about 4 months. During the test, the system automatically collected and stored the data of temperature, weight, and climatic conditions throughout the process. Table 3 shows the average climatic conditions at each stage of the test process. When leaving the silo, samples were taken at each testing point according to the placement of the silo and the height of the grain layer. During the sampling process, the original weight of the silo and the thickness of the grain layer were recorded first, and the corn was sorted out layer by layer starting from the top to the bottom until it reached the temperature sensor position of each layer and then sampled. After reaching the temperature sensor position, the mass of the storage silo (including the remaining corn) and the corresponding thickness of the grain layer were weighed again. When sampling, 8–10 ears of corn were taken as samples near each temperature sensor in the order from west to east. After threshing, samples were put it into the sample bags prepared in advance. The samples were divided into two bags at each sampling point, with each bag containing not less than 500 g. During the sampling process, the remaining corn samples should be kept at the same level as the temperature sensor.

5.4. Calculation of the Absolute Water Potential of the Grain Pile

The absolute water potentials of air and grain were used to calculate the absolute water potential of corn during natural ventilation and the absolute water potential of air under experimental conditions, respectively, to assess the absolute water potential variation and distribution of grain piles and the drying effect of grain silos.

$$E_{ja} = 8.31 \times (t_a + 273) \times \ln(100 \times \exp(\frac{87.72 \times \lg(RH_a) + 0.9845 \times (1737.1 - \frac{474242}{273+t_g}) - 270.57}{87.72}) \times 133.3)/18 \quad (1)$$

$$E_{jg} = 8.31 \times (t_g + 273) \times \ln\left(\exp\left(\frac{\left(\frac{D_0}{222} \times \left(e^{\frac{B_1-M}{A_1}} - e^{\frac{B_2-M}{A_2}}\right) + 0.9845\right) \times \left(1737.1 - \frac{474242}{273+t_g}\right) + D_0 \times \left(1 - e^{\frac{B_1-M}{A_1}}\right) - 68.57}{87.72}\right) \times 133.3\right)/18 \qquad (2)$$

where E_{ja} was the absolute water potential of air, kJ/kg; E_{jg} was the absolute water potential of grain, kJ/kg; M was the wet basis moisture content of grain, %; t_g was the temperature of grain, °C; RH_a was the relative humidity of air, %; t_a was the temperature of air, °C; and A_1, A_2, B_1, B_2, and D_0 were the desorption parameters of corn, 4.393, 4.845, 7.843, 3.858, and 203.892, respectively.

5.5. Determination of Aflatoxin Content

The aflatoxin content in food was determined regarding the Chinese national standard GB 5009.22–2016. The sample was extracted with methanol–water, the extract was filtered and diluted, and the filtrate was purified by immunoaffinity chromatography containing aflatoxin-specific antibodies. This antibody was specific for aflatoxin B1, B2, G1, and G2. Aflatoxin was cross-linked to the antibody in the chromatography medium. Impurities were removed from the immunoaffinity column with water or Tween-20/phosphate-buffered solution (PBS). Elution was performed with methanol through the immunoaffinity chromatography column. The eluate was passed through the column of HPLC. It was then derivatized using an AURA photochemical derivatization cell and detected by a fluorescence detector and quantified by external standard method.

5.6. Determination of Deoxynivalenol Content

The determination of deoxynivalenol in food was carried out regarding the Chinese national standard GB 5009.111–2016. Deoxynivalenol was extracted from the sample. After purification and concentration by immunoaffinity column, the sample was determined by HPLC with UV detector and quantified by external standard method.

5.7. Determination of Zearalenone Content

The determination of zearalenone in cereals was carried out regarding the national standard GB 5009.209–2016. Zearalenone was extracted from the sample with acetonitrile–water, and the extract was cleaned up and concentrated by an immunoaffinity column. The determination was performed by HPLC with a fluorescence detector and quantified by external standard method.

The method validation involved in the above six mycotoxin assays is shown in Table 21.

Table 21. Summary of the method validation.

Mycotoxin [a]	AFB1	AFB2	AFG1	AFG2	ZEN	DON
Coefficient of correlation (R^2)	0.999	0.999	1.000	0.999	0.999	0.999
Range (ng/mL)	0.1–40	0.03–12	0.1–40	0.03–12	10–500	100–5000
Spiked level (µg/kg)	10	10	10	10	30.0	300.0
Recovery (%)	93.7	105.3	96.9	101.1	97.5	96.6
RSD (%) *	10.2	17.8	11.4	16.1	2.7	2.2
LOD (µg/kg)	0.03	0.01	0.03	0.01	5	100
LOQ (µg/kg)	0.1	0.03	0.1	0.03	17	200

Note: [a] AFB1, AFB2, AFG1, AFG2: aflatoxins B1, B2, G1, and G2; ZEN: zearalenone; DON: deoxynivalenol. RSD: relative standard deviation; LOD: limit of detection; LOQ: limit of quantification. * This value was calculated from 3 replicates of mycotoxin analysis.

5.8. Mold Rate

Chinese national standard GB1353-2009 was used for the mold rate to make clear provisions—raw mold particles need to be $\leq 2\%$.

5.9. Statistical Analysis

Statistical analysis was performed using Origin 9.0, and data were expressed as mean standard deviation SD (n = 3). Significant differences between means (p < 0.05) were investigated by Tukey's test, using one-way ANOVA with SPSS 17.0.

Author Contributions: Conceptualization, C.S. and W.W.; methodology, C.S. and W.W.; formal analysis, T.H.; investigation, Y.X.; data curation, T.H.; writing—original draft preparation, J.Z.; writing—review and editing, J.Z. and Y.X.; supervision, W.W.; project administration, Y.X. and W.W. All authors have read and agreed to the published version of the manuscript.

Funding: This research was funded by the project "Key technology development for food biological control", project number 2016413.

Institutional Review Board Statement: Not applicable.

Informed Consent Statement: Not applicable.

Data Availability Statement: Data is contained within this article.

Conflicts of Interest: The authors declare no conflict of interest. The funders had no role in the design of the study; in the collection, analyses, or interpretation of data; in the writing of the manuscript, or in the decision to publish the results.

References

1. Rouf Shah, T.; Prasad, K.; Kumar, P. Maize–A Potential Source of Human Nutrition and Health: A Review. *Cogent Food Agr.* **2016**, *2*, 1166995. [CrossRef]
2. Chulze, S.N. Strategies to Reduce Mycotoxin Levels in Maize during Storage: A Review. *Food Addit. Contam. Part A* **2010**, *27*, 651–657. [CrossRef] [PubMed]
3. Neme, K.; Mohammed, A. Mycotoxin Occurrence in Grains and the Role of Postharvest Management as a Mitigation Strategies. A Review. *Food Control* **2017**, *78*, 412–425. [CrossRef]
4. Palumbo, R.; Gonçalves, A.; Gkrillas, A.; Logrieco, A.; Dorne, J.-L.; Dall'Asta, C. Mycotoxins in Maize: Mitigation Actions, with a Chain Management Approach. *Phytopathol. Mediterr.* **2020**, *59*, 5–28. [CrossRef]
5. Hu, L.; Liu, H.; Yang, J.; Wang, C.; Wang, Y.; Yang, Y.; Chen, X. Free and Hidden Fumonisins in Raw Maize and Maize–Based Products from China. *Food Addit. Contam. Part B* **2019**, *12*, 90–96. [CrossRef] [PubMed]
6. Schaarschmidt, S.; Fauhl-Hassek, C. The Fate of Mycotoxins during the Primary Food Processing of Maize. *Food Control* **2021**, *121*, 107651. [CrossRef]
7. Sun, X.D.; Su, P.; Shan, H. Mycotoxin Contamination of Maize in China. *Compr. Rev. Food Sci. Food Saf.* **2017**, *16*, 835–849. [CrossRef]
8. Mukanga, M.; Derera, J.; Tongoona, P.; Laing, M.D. A Survey of Pre-Harvest Ear Rot Diseases of Maize and Associated Mycotoxins in South and Central Zambia. *Int. J. Food Microbiol.* **2010**, *141*, 213–221. [CrossRef]
9. Affognon, H.; Mutungi, C.; Sanginga, P.; Borgemeister, C. Unpacking Postharvest Losses in Sub-Saharan Africa: A Meta-Analysis. *World Dev.* **2015**, *66*, 49–68. [CrossRef]
10. Kaminski, J. Post-Harvest Loss in Sub–Saharan Africa—What Do Farmers Say? *Glob. Food Secur.* **2014**, *3*, 149–158. [CrossRef]
11. Omotilewa, O.J.; Ricker-Gilbert, J.; Ainembabazi, J.H.; Shively, G.E. Does Improved Storage Technology Promote Modern Input Use and Food Security? Evidence from a Randomized Trial in Uganda. *J. Dev. Econ.* **2018**, *135*, 176–198. [CrossRef] [PubMed]
12. Tefera, T. Post-Harvest Losses in African Maize in the Face of Increasing Food Shortage. *Food Secur.* **2012**, *4*, 267–277. [CrossRef]
13. Gitonga, Z.M.; De Groote, H.; Kassie, M.; Tefera, T. Impact of Metal Silos on Households' Maize Storage, Storage Losses and Food Security: An Application of a Propensity Score Matching. *Food Policy* **2013**, *43*, 44–55. [CrossRef]
14. James, A.; Zikankuba, V.L. Mycotoxins Contamination in Maize Alarms Food Safety in Sub-Sahara Africa. *Food Control* **2018**, *90*, 372–381. [CrossRef]
15. Yuya, A.I.; Tadesse, A.; Azerefegne, F.; Tefera, T. Efficacy of Combining Niger Seed Oil with Malathion 5% Dust Formulation on Maize against the Maize Weevil, *Sitophilus Zeamais* (Coleoptera: Curculionidae). *J. Stored Prod. Res.* **2009**, *45*, 67–70. [CrossRef]
16. Li, F. Vista of the farmer's grain storage facilities in China. *Sci. Technol. Cereals Oils Foods* **2012**, *20*, 50–52. (In Chinese)
17. Luo, Y.; Wu, L. Farm corn storage losses in different scales and its main determinants. *J. Maize Sci.* **2021**, *29*, 177–183. (In Chinese)
18. Bwambale, J.; Durodola, O.S.; Nabunya, V. Development and Evaluation of an Improved Maize Silo to Advance Food Security in Uganda. *Cogent Food Agric.* **2020**, *6*, 1834666. [CrossRef]
19. Tefera, T.; Kanampiu, F.; De Groote, H.; Hellin, J.; Mugo, S.; Kimenju, S.; Beyene, Y.; Boddupalli, P.M.; Shiferaw, B.; Banziger, M. The Metal Silo: An Effective Grain Storage Technology for Reducing Post-Harvest Insect and Pathogen Losses in Maize While Improving Smallholder Farmers' Food Security in Developing Countries. *Crop Prot.* **2011**, *30*, 240–245. [CrossRef]
20. Chuma, T.; Mudhara, M.; Govereh, J. Factors Determining Smallholder Farmers' Willingness to Pay for a Metal Silo in Zimbabwe. *Agrekon* **2020**, *59*, 202–217. [CrossRef]

21. De Groote, H.; Kimenju, S.C.; Likhayo, P.; Kanampiu, F.; Tefera, T.; Hellin, J. Effectiveness of Hermetic Systems in Controlling Maize Storage Pests in Kenya. *J. Stored Prod. Res.* **2013**, *53*, 27–36. [CrossRef]
22. Njoroge, A.W.; Affognon, H.D.; Mutungi, C.M.; Manono, J.; Lamuka, P.O.; Murdock, L.L. Triple Bag Hermetic Storage Delivers a Lethal Punch to *Prostephanus Truncatus* (Horn) (Coleoptera: *Bostrichidae*) in Stored Maize. *J. Stored Prod. Res.* **2014**, *58*, 12–19. [CrossRef]
23. Baoua, I.B.; Amadou, L.; Ousmane, B.; Baributsa, D.; Murdock, L.L. PICS Bags for Post-Harvest Storage of Maize Grain in West Africa. *J. Stored Prod. Res.* **2014**, *58*, 20–28. [CrossRef]
24. García–Lara, S.; García–Jaimes, E.; Ortíz–Islas, S. Field Effectiveness of Improved Hermetic Storage Technologies on Maize Grain Quality in Central Mexico. *J. Stored Prod. Res.* **2020**, *87*, 101585. [CrossRef]
25. Williams, S.B.; Murdock, L.L.; Baributsa, D. Safe Storage of Maize in Alternative Hermetic Containers. *J. Stored Prod. Res.* **2017**, *71*, 125–129. [CrossRef]
26. Walker, S.; Jaime, R.; Kagot, V.; Probst, C. Comparative Effects of Hermetic and Traditional Storage Devices on Maize Grain: Mycotoxin Development, Insect Infestation and Grain Quality. *J. Stored Prod. Res.* **2018**, *77*, 34–44. [CrossRef]
27. Murdock, L.L.; Margam, V.; Baoua, I.; Balfe, S.; Shade, R.E. Death by Desiccation: Effects of Hermetic Storage on Cowpea Bruchids. *J. Stored Prod. Res.* **2012**, *49*, 166–170. [CrossRef]
28. Liu, B.; Hu, T. Study on the design of the corn ear storage Bin. *Modern Food* **2016**, *22*, 126–128. (In Chinese)
29. Dong, D.; Li, J.; Dong, M.; Lin, L.; Gao, S.; Zhou, Y.; Wang, D. Test of granary loaded with corn ear for farmers in Northest China. *Sci. Technol. Cereal Oils Food* **2015**, *23*, 97–99. (In Chinese)
30. Dong, D.; Zhou, Y. Experiment and discussion of drying corn ears in the silo. *Sci. Technol. Cereal Oils Food* **2010**, *18*, 47–48. (In Chinese)
31. Wu, F.; Yan, X.; Yang, Y. Survey of aflatoxin B_1 contamination in maize stored by farmers in China in 2016. *Cereal Feed Ind.* **2018**, *1*, 8–12.
32. Wu, F.; Zhu, Y.; Yan, X.; Yang, Y. A review of reducing grain storage lossed of farmers in developing countries. *Grain Storage* **2018**, *6*, 15–24. (In Chinese)
33. Yin, L.; Yu, F.; Wu, W. Exploring the mechanism of low–temperature vacuum drying for cereal grain. *J. Chin. Cereals Oils Assoc.* **2006**, *21*, 129–132. (In Chinese)
34. Yin, L. *Research of Heat and Mass Transfer Model and Dielectric Properties of Maize under Vacuum Drying Conditions Based on Water Potential*; Jilin University: Changchun, China, 2011. (In Chinese)
35. Wu, Z.; Wu, W.; Chen, S.; Liu, Z. A Method for Managing and Controlling Mechanical Ventilation Operations in Grain Silos Using Water Potential Map. Chinese Patent 201410265033.9, 25 January 2017.
36. Liu, Z.; Wu, Z.; Wang, X.; Song, J.; Wu, W. Numerical simulation and experimental study of deep bed corn drying based on water potential. *Math. Probl. Eng.* **2015**, *2015*, 539846. [CrossRef]
37. Wu, Z.; Li, X.; Guo, D.; Wang, S. The use of a moisture sorption equation for depot aeration of shelled corn in China. In Proceedings of the 2011 CIGR Section VI International Symposium on Towards a Sustainable Food Chain Food Process, Bioprocessing and Food Quality Management, Nantes, France, 18–20 April 2011.
38. Li, X.; Wei, Z.; Cao, Z.; Feng, Q.; Wang, J. Equilibrium moisture content and sorption isosteric heats of five wheat varieties in China. *J. Stored Prod. Res.* **2011**, *47*, 39–47. [CrossRef]
39. Xiong, Q.; Wei, H.; Gao, L.; Wang, P.; Lin, D.; Guo, F. Research methods and progress on the flight behavior of stored grain pests. *Grain Sci. Technol. Econ.* **2018**, *43*, 63–67. (In Chinese)
40. Luo, F.; Lu, X.; Guo, J. Correlation analysis of corn moldy kernels, mildewed kernels and mycotoxin content. *Grain Oils Storage Technol. Newsl.* **2020**, *36*, 42–44. (In Chinese)
41. Chen, S.; Wu, W.; Xu, Y.; Han, F.; Qin, X. Dynamics analysis on water migration of corn ear during natural ventilation. *Trans. Chin. Soc. Agr. Eng.* **2016**, *32*, 277–282. (In Chinese)

Article

Detection of αB-Conotoxin VxXXIVA (αB-CTX) by ic-ELISA Based on an Epitope-Specific Monoclonal Antibody

Hengkun Tang [†], Haimei Liu [†], Yehong Gao, Rui Chen, Mingke Dong, Sumei Ling, Rongzhi Wang * and Shihua Wang *

Key Laboratory of Pathogenic Fungi and Mycotoxins of Fujian Province, Key Laboratory of Biopesticide and Chemical Biology of Education Ministry, School of Life Sciences, Fujian Agriculture and Forestry University, Fuzhou 350002, China; tanghengkun@fafu.edu.cn (H.T.); wuhuijuan@fafu.edu.cn (H.L.); gaoyehong@fafu.edu.cn (Y.G.); chenrui@fafu.edu.cn (R.C.); dmkhubei@fafu.edu.cn (M.D.); lsmpu2008@m.fafu.edu.cn (S.L.)
* Correspondence: wrz0629@fafu.edu.cn (R.W.); wshmail@m.fafu.edu.cn (S.W.)
† These authors contributed to this work equally.

Abstract: In view of the toxicological hazard and important applications in analgesics and cancer chemotherapeutics of αB-CTX, it is urgent to develop an accurate, effective and feasible immunoassay for the determination and analysis of αB-CTX in real samples. In this study, MBP-αB-CTX4 tandem fusion protein was used as an immunogen to elicit a strong immune response, and a hybridoma cell 5E4 secreting IgG2b against αB-CTX was successfully screened by hybridoma technology. The affinity of the purified 5E4 monoclonal antibody (mAb) was 1.02×10^8 L/mol, which showed high affinity and specificity to αB-CTX. Epitope 1 of αB-CTX is the major binding region for 5E4 mAb recognization, and two amino acid residues (14L and 15F) in αB-CTX were critical sites for the interaction between αB-CTX and 5E4 mAb. Indirect competitive ELISA (ic-ELISA) based on 5E4 mAb was developed to detect and analyze αB-CTX in real samples, and the linear range of ic-ELISA to αB-CTX was 117–3798 ng/mL, with a limit of detection (LOD) of 81 ng/mL. All the above results indicated that the developed ic-ELISA had high accuracy and repeatability, and it could be applied for αB-CTX detection and drug analysis in real samples.

Keywords: αB-conotoxin VxXXIVA; epitope; hybridoma; monoclonal antibody; ELISA

Key Contribution: To our knowledge, this is the first report on the preparation of a specific mAb against αB-CTX and the development of ic-ELISA to determine and analyze the content of αB-CTX in real samples. αB-CTX (1–20) is the major binding region of αB-CTX in mAb 5E4 recognition, and amino acids 14L and 15F were two critical sites for the interaction between mAb 5E4 and αB-CTX. In view of the excellent signal stability, high repeatability and accuracy of detection method, the developed ic-ELISA in this study could be used for drug identification and rapid detection of αB-CTX in real seafood samples.

Citation: Tang, H.; Liu, H.; Gao, Y.; Chen, R.; Dong, M.; Ling, S.; Wang, R.; Wang, S. Detection of αB-Conotoxin VxXXIVA (αB-CTX) by ic-ELISA Based on an Epitope-Specific Monoclonal Antibody. *Toxins* **2022**, *14*, 166. https://doi.org/10.3390/toxins14030166

Received: 9 January 2022
Accepted: 21 February 2022
Published: 23 February 2022

1. Introduction

Recently, cone snails have been regarded as one of the focuses for new drug discovery [1], with the majority of their distribution throughout tropical and subtropical waters, such as the South China Sea, Australia, and the Pacific Ocean [1,2]. Conotoxins, a diverse array of unique bioactive neurotoxins, are mainly secreted by different kinds of cone snails [3,4]. Conotoxins are often toxic to humans and animals and can cause convulsions and paralysis, even to death [5]. More than 80,000 natural conotoxins have been estimated to exist in various cone snails around the world [6–8], and all conotoxins can be divided into 26 gene superfamilies based on their conserved signal sequences and their characteristic cysteine framework [5,9]. Among them, α-, μ- and ω-conotoxins are the most characterized families so far for their high specificity and affinity to ion channels [1,10–12].

αB-conotoxin VxXXIVA (αB-CTX) is a new conotoxin found in *Conus vexillum*, which belongs to the B-gene superfamily. It is reported that the precursor of αB-CTX is different from other conotoxins as it lacks a pro region responsible for enhancing oxidative folding and secretion of hydrophobic O-superfamily conotoxins [12]. Specifically, such a mechanism is not necessary for the more hydrophilic αB-CTX. The mature αB-CTX is a peptide with 40 amino acid residues containing only 4 cysteine residues, and the most active disulphide linkages of αB-CTX are C-CC-C [12]. αB-CTX is a specific nicotinic acetylcholine receptor (nAChR) antagonist with the greatest potency against the α9α10 subtype [12]. The α9α10 nAChR is an important target for the development of analgesics and cancer chemotherapeutics [13–16], and αB-CTX represents a novel ligand with which to probe the structure and function of this protein. Given the toxicological effects and potential application value of αB-CTX, it is urgent to develop an accurate and sensitive method to identify and detect αB-CTX in sea samples.

At present, the main analysis and detection methods for αB-CTX are analytical chemistry, including high-performance liquid chromatography (HPLC), nuclear magnetic resonance (NMR) and circular dichroism (CD) [12]. Although these methods have high sensitivity and accuracy for analysis of the content and composition of αB-CTX, the main disadvantages of these methods are that they are time-consuming, require complex sample pre-treatment and require expensive equipment and trained professionals [17,18]. Moreover, complicated and tedious operation steps, high requirement of samples and detection practicability seriously limits their application in the detection of αB-CTX in real samples [19]. In contrast, immunoassays with high sensitivity, accuracy and adaptability have been widely used to detect the targets (toxins, pathogenic molecules and heavy metals) in different samples [20–22]. In the previous study, we developed ELISA and colloidal test strips based on monoclonal antibodies (mAb) against ω-CTX to detect ω-CTX MVIIA residue in Conus samples [5]. Specifically, no related immunoassay was reported to detect and analyze αB-CTX in real samples until now. This is may be due to the difficulty in acquiring antigens or the low immunogenicity of antigens, which fail to induce a strong immune response for antibody production. Fortunately, the antigen epitope analysis and antigen tandem fusion expression established in this study can effectively solve these key problems and can be used to solve these important limits. Therefore, the aim of this study was to design and prepare an antigen with strong immunogenicity to elicit the antibody production in mouse models and to screen an epitope-specific monoclonal antibody by hybridoma technology. The key binding regions and sites between the resulted mAb and αB-CTX were further analyzed. Finally, ic-ELISA based on epitope-specific mAb was established and used to detect and analyze the content of αB-CTX toxin in actual samples.

2. Results and Discussion

2.1. Epitope Binding Pattern Evaluation

αB-CTX is a peptide consisting of 40 amino acid residues in length. The antigenic sites on αB-CTX were first analyzed using the Bepipred B-cell linear epitope prediction tool [23]. As shown in Figure 1, this antigen contains two major epitopes located at positions 6–15 and 19–27 in the protein sequence, and the regions of the predicted antigenic sequences were designated as EP1 and EP2, respectively (Figure 1A,B). Interestingly, the αB-CTX has four Cys residues on its peptide sequence, and the specific disulfide bond arrangement of this peptide is C-CC-C (Figure 1C, chocolate yellow) [12]. Meanwhile, the predicted 3D model of αB-CTX resembles a triangular clamp, with a helix at each end. The key binding sites (Leu14 and Phe15) between antibody and antigen interactions were marked as different colors in the predicted 3D model and linear amino acid sequences (Figure 1D, dark blue and red).

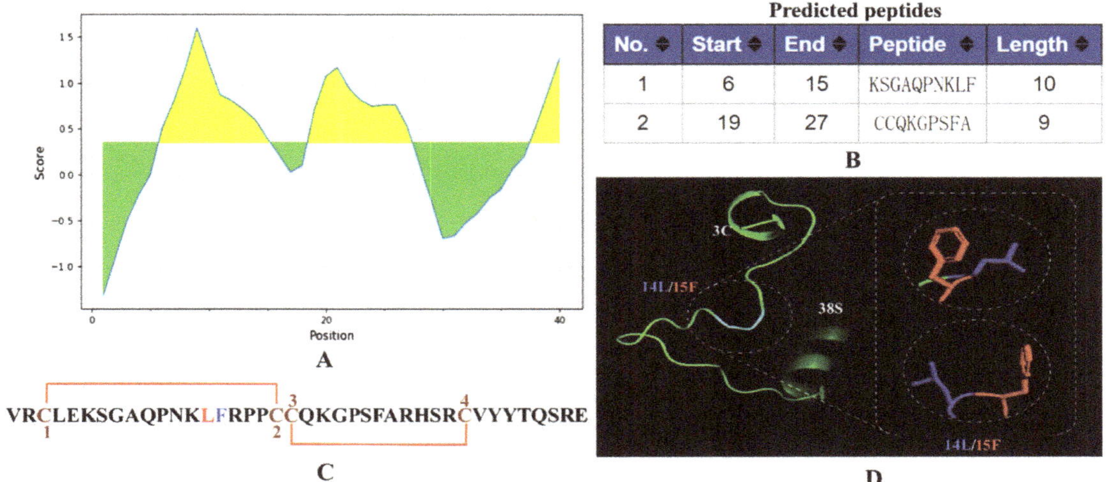

Figure 1. Antigen epitope prediction and 3D structure analysis. (**A**) Output graph of the predicted B-cell linear epitopes showing the amino acid position (x-axis) and Bepipred score (Y-axis). Residues (yellow) with scores above the default threshold of 0.35 have a higher probability of being part of an epitope. The green indicates lower probability of being part of an epitope. (**B**) Two predicted epitopes of lengths (start and end positions were indicated; 1: EP1 (6–15); 2: EP2: (19–27)). (**C**) Sequence and disulfide connectivity of αB-CTX. (**D**) Swiss model of αB-CTX from Cys3 to Ser38 using Leishmania Major mitochondrial ribosome (PDB entry 7ane.36.A) as template. Amino acid residues Leu14 and Phe15 are shown in dark blue and red, respectively. The sequences upstream of Cys3 and downstream of Ser38 were not modeled.

2.2. Preparation of αB-CTX and Animal Immunization

The connected tetramer fragments (αB-CTX4) and αB-CTX monomer were used to construct three different expression vectors for protein expression and purification (Figure 2A). The fusion protein MBP-LK-αB-CTX4 and the two other fusions (TRX-αB-CTX and GST-αB-CTX) were shown in Figure 2B,C. These fusion proteins were successfully expressed after IPTG induction and further purified by Ni^{2+}-NTA affinity chromatography. The band sizes of the purified MBP-LK-αB-CTX4 and TRX-αB-CTX with high purity on the gel of SDS-PAGE were 60 kDa and 28 kDa, respectively, the same as the theoretical result [5]. Then, the purified MBP-LK-αB-CTX4 was used as the immunogen for animal immunization, and the titer of serum was tested by ELISA with the TRX-αB-CTX as the detection antigen (Figure 2D). Compared to the immunized mouse 2 and control, the immunized mouse 1 had the highest antiserum titer to the target antigen (reaching 1:16,000), indicating that a strong immune response was induced by the MBP-LK-αB-CTX4 antigen.

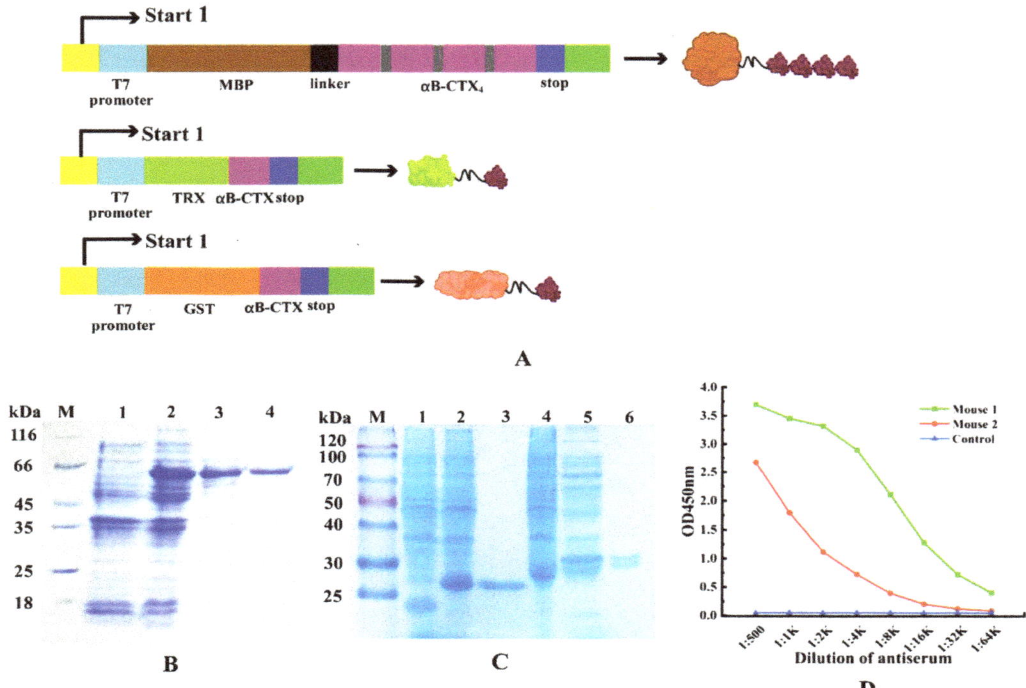

Figure 2. Purification of fusion proteins and animal immunization. (**A**) Construction of fusion protein expression vectors (MBP-LK-αB-CTX4, TRX-αB-CTX and GST-αB-CTX). (**B**) Expression and purification of fusion proteins MBP-LK-αB-CTX4. Lane M: middle molecular protein marker, Lane 1: the expressed product of pBD-*mbp*/BL21(DE3), Lane 2: the expressed total proteins of pBD-*mbp-lk-αB-CTX4*/BL21(DE3), Lane 3–4: the purified product of MBP-LK-αB-CTX4. (**C**) Expression and purification of fusion proteins TRX-αB-CTX and GST-αB-CTX. Lane M: middle molecular protein marker, Lane 1: the expressed product of pET-32a/BL21(DE3), Lane 2: the expressed total proteins of pET-32a-*αB-CTX*/BL21(DE3), Lane 3: the purified product of TRX-αB-CTX, Lane 4: the expressed product of pGEX-6P-1/BL21(DE3), Lane 5: total proteins of pGEX-6P-1-*αB-CTX*/BL21(DE3), Lane 6: the purified product of GST-αB-CTX. (**D**) Titer determination of the immunized mouse by ELISA.

2.3. Screening and Identification of Positive Hybridoma

After fusion, the cells in the 96-well plate were composed of 3 kinds of cells: myeloma cells, mouse spleen cells and feeder layer cells. These cells were cultured in a selective medium containing 2% HAT and 20% FBS [20]. From the third day, the fusion cell colonies could be obviously observed and gradually grown as time went on, and most of the unfused cells were apoptotic (Figure 3A). On the 8th day after fusion, the supernatant of the culture was taken to detect whether the cells secreted antibodies by ELISA, and 5 positive hybridomas were screened out successfully. Combined with the growth status of cells and antibody properties, the hybridoma 5E4 was finally selected for subsequent experiments (Figure 3B). As shown in Figure 3C, the IgG2b subtype has the highest value, while the values of other types are very low, further indicating that the subtype of the 5E4 cell line belongs to IgG2b. The chromosome analysis result of hybridoma 5E4 in Figure 3D showed that the chromosome numbers of hybridoma 5E4 were 106 ± 2, corresponding to the sum of the chromosome number of myeloma cells and mouse spleen cells in theory [21]. The ideal hybridoma should be selected based on multi-factors. Although the supernatant titer of hybridoma 5E4 was not the best in Figure 3B, the antibody secreted by 5E4 exhibited the best specificity and stability to αB-CTX compared to other hybridomas. Hence, it was chosen as the ideal cell line for further antibody purification.

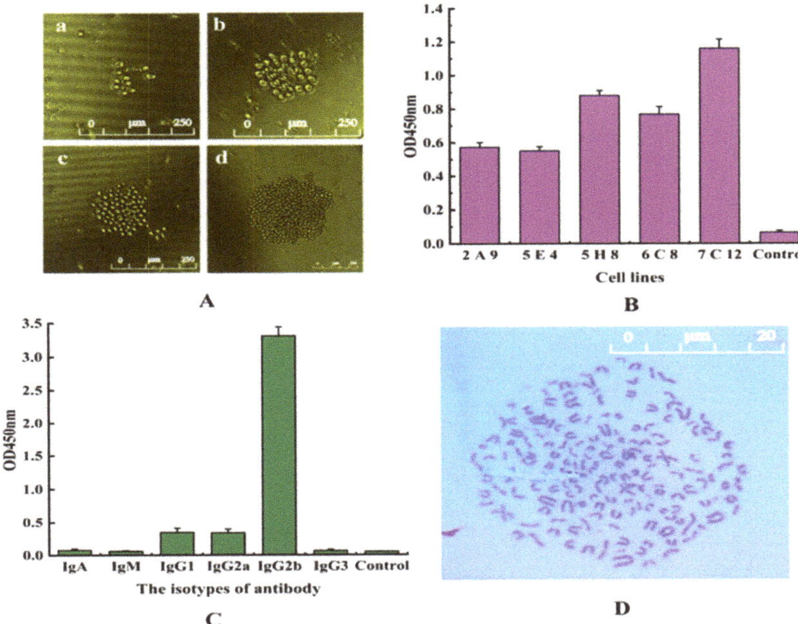

Figure 3. Hybridoma screening and characterization. (**A**) Hybridoma cell culture after fusion. a: the 3rd day, b: the 5th day, c: the 7th day, d: the 9th day. (**B**) ELISA assay of the selected positive hybridoma cell lines. Five positive hybridoma cell lines were obtained and named 2E9, 5E4, 5H8, 6C8 and 7C12. (**C**) Isotyping of 5E4 cells secreting anti-αB-CTX mAb by using an isotyping kit. (**D**) Analysis of the chromosome of 5E4 cell line.

2.4. Purification and Identification of 5E4 mAb

To obtain the monoclonal antibody 5E4 with high affinity and purity, the cultured hybridoma 5E4 cells were injected into the abdominal cavity of pre-sensitized mice to produce ascites, and the anti-αB-CTX4 mAb was purified by Protein G affinity chromatography. SDS-PAGE results, as shown in Figure 4A, indicated that the purified antibody has two protein bands with molecular weights of 50 kDa and 25 kDa on gel, corresponding to the heavy chain and light chain, respectively. The titer of ascites and purified antibody were shown in Figure 4B, and the purified 5E4 mAb exhibited similar αB-CTX binding activity to that of ascites. In this experiment, the affinity of the resulting 5E4 mAb was determined by indirect ELISA (iELISA), and the affinity curve was drawn using Origin 8.0 software. The results, shown in Figure 4C demonstrated that the affinity of 5E4 mAb was 1.02×10^8 L/mol, belonging to high affinity antibody. Meanwhile, the specificity of 5E4 mAb was further analyzed, and the result was shown in Figure 4D. As expected, the prepared 5E4 mAb could specifically bind to the two different fusion proteins GST-αB-CTX and TRX-αB-CTX in ELISA assay, but not react with the other antigens, such as GST-μ-CTX, TRX-μ-CTX, TRX-ω-CTX, SN311 and SN285, indicating that this 5E4 mAb with high affinity was specific to αB-CTX.

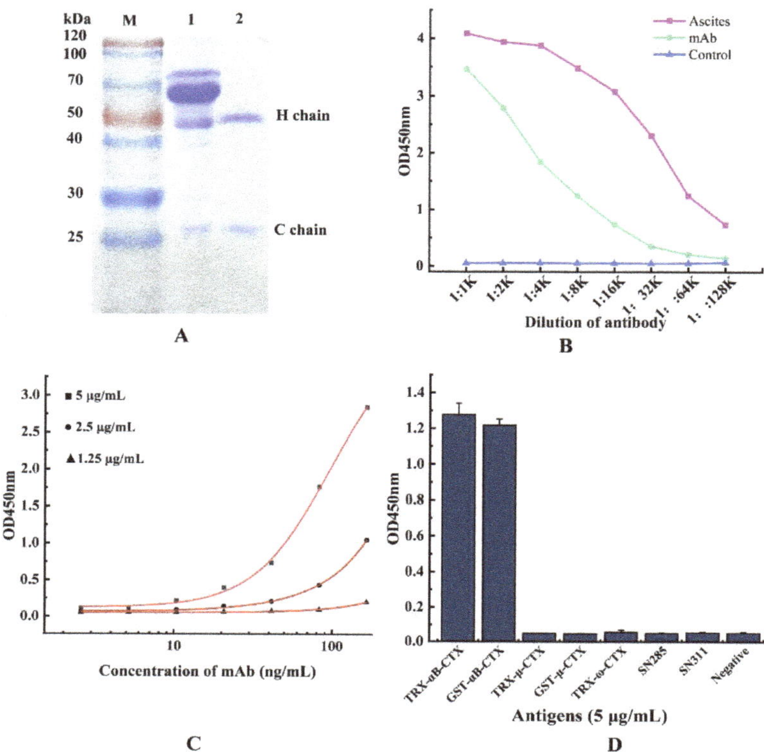

Figure 4. Purification and characterization of anti-αB-CTX mAb. (**A**) SDS-PAGE analysis of the purified mAb. Lane M: standard protein marker, Lane 1: the total protein of ascites, Lane 2: the purified anti-αB-CTX mAb. (**B**) The titer of ascites and anti-αB-CTX mAb. (**C**) Affinity determination of anti-αB-CTX mAb by iELISA. The affinity constant of the αB-CTX mAb is 1.02×10^8 L/mol. (**D**) The specificity of anti-αB-CTX mAb was determined by ic-ELISA.

2.5. αB-CTX (1–20) Is the Major Binding Region by 5E4 mAb Recognition

After the preparation of the antibody, how to investigate the binding area for 5E4 mAb-αB-CTX interactions became one of the main concerns of this study. To achieve this purpose, full-length αB-CTX and three different αB-CTX fragments including αB-CTX (1–20), αB-CTX (10–30) and αB-CTX (20–40) were designed and synthesized (Figure 5A), and ELISA was used to test the binding activity of these fragments. The ELISA results in Figure 5C showed that the purified 5E4 mAb has a strong binding signal to αB-CTX (1–20), similar to that of full-length αB-CTX, with the binding activity being far higher than others. Compared to αB-CTX (1–20), αB-CTX (10–30) and αB-CTX (20–40) are not recognized by the 5E4 mAb. The result derived from the ELISA indicated that αB-CTX (1–20) is the major binding region for 5E4 mAb recognition. Interestingly, the predicted first epitope 1 (EP1) of αB-CTX was just located in the region of 6–15 (Figure 1B), further suggesting that the EP1 may be the major region for 5E4 mAb recognition.

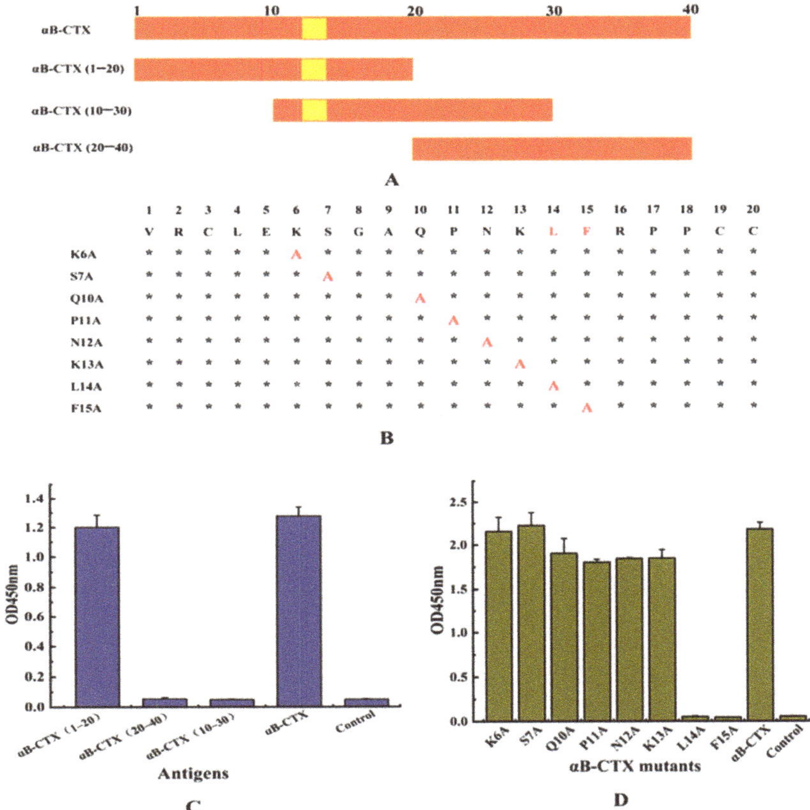

Figure 5. Epitope identification for 5E4 anti-αB-CTX mAb interactions. (**A**) Three regions of αB-CTX, including potential binding sites (Val1-Cys20, Gln10-Ser30 and Cys20-Glu40). (**B**) Epitope mapping of region Val1-Cys20 by alanine-scan mutagenesis of the mAb binding site. Critical residues (highlighted in red) are defined as those with alanine substitution. *: the unmutated amino acid site; A: the amino acid was mutated to alanine. (**C**) The titer for various peptide regions and αB-CTX full-length peptide. (**D**) The titer for various site mutants and αB-CTX full-length peptide.

2.6. L14 and F15 Are the Critical Sites between 5E4 mAb-αB-CTX Interactions

To further analyze the critical sites responsible for 5E4 mAb-αB-CTX interactions, 8 amino acids located at the epitope 1 (6–15) of αB-CTX, except for G8 and A9, were mutated to alanine by site-directed mutagenesis (Figure 5B). The ELISA result in Figure 5D showed that the binding activity of 5E4 mAb against 2 mutated amino acids (L14A and F15A) was all obviously decreased, similar to the negative control. Other mutants (K6A, S7A, Q10A, P11A, N12A and K13A) exhibited strong binding signals to 5E4 mAb recognition as the maternal αB-CTX. The above results demonstrated that amino acids L14 and F15 located at the epitope 1 of αB-CTX were critical sites for the interaction between 5E4 mAb and αB-CTX.

2.7. Establishment of ic-ELISA for αB-CTX

αB-CTX is a hapten with low molecular weight and limited epitopes, so it is very difficult to develop a double antibody sandwich ELISA for αB-CTX detection and analysis. Combining the situation of 5E4 mAb and αB-CTX, indirect competitive ELISA (ic-ELISA) was applied to detect αB-CTX in real samples after optimizing the antigen and antibody concentration. As shown in Figure 6, the typical calibration curve was drawn by plotting (B/B0) against αB-CTX concentration, and the equation of the logistic curve was $y = 0.0728 + [(1.04855 - 0.0728)/(1 + x/504.72267)^{0.93828}]$, with a correlation coefficient (R^2) of 0.98944 (Figure 6A). From the result shown in Figure 6B, the linear equation was $y = 1.97071 - 0.51887x$, and the correlation coefficient (R^2) was about 0.97819. In ic-ELISA, the limit of detection (LOD) for αB-CTX was 81 ng/mL, which is defined as the concentration of target antigen corresponding to the inhibition rate that reached 10% [19,24]. Furthermore, the linear range of detection was 117~3798 ng/mL with a half inhibitory concentration (IC50) of 661 ng/mL.

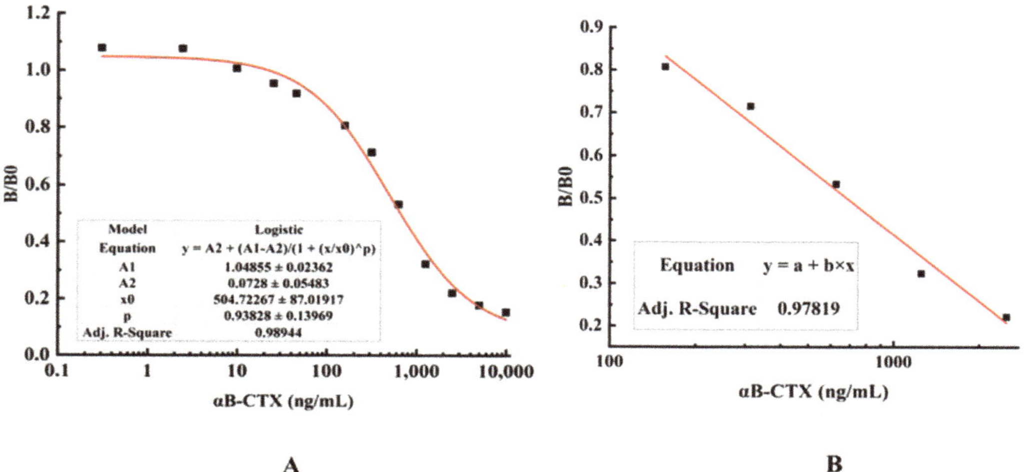

Figure 6. Standard curves for αB-CTX detection. (**A**) A typical calibration curve illustrated by plotting (B/B0) against αB-CTX. The data obtained with various inhibitor concentrations and without inhibitor are referred to as B and B0, respectively. The equation is $y = 0.0728 + (1.04855 - 0.0728)/((1 + (x/504.72267)^{0.93828})$, with a correlation coefficient (R^2) of 0.98944. The limit of detection (LOD) was 81 ng/mL for αB-CTX. (**B**) The linear portion of the standard curve. The equation is $y = 1.97071 - 0.51887x$, with a correlation coefficient (R^2) of 0.97819. The range of linear detection for αB-CTX was 117–3798 ng/mL.

2.8. Detection of αB-CTX by ic-ELISA in Spiked and Real Samples

In this study, the recovery and variation coefficients for αB-CTX detection were calculated by ic-ELISA, and the PBSM solution without any contamination was spiked with different concentrations of αB-CTX [18]. All analyses were performed with three parallel experiments. As shown in Table 1, the recovery of detection ranged from (86.62 ± 0.72)% to (96.46 ± 6.09)%, with an average of (92.93 ± 3.69)%, and the average variation coefficient (CV) was 3.87% in intra-assay. The average recovery and CV in inter-assay were (96.31 ± 4.48)% and 4.52%, respectively. The results from ELISA detection indicated that the developed ic-ELISA had high accuracy and repeatability, and it could be used to detect the content of αB-CTX in real samples. Finally, five different shellfish seafood products were purchased randomly from local supermarkets as experimental detection materials by ic-ELISA. The detection results in Table 2 showed that the values of these real samples were consistent with that of PBS, indicating that none of these actual samples contained α-CTX toxin.

Table 1. The determination of recovery from spiked samples.

Spiked Level (ng/mL)	Intra-Assay				Inter-Assay			
	n	Measured (ng/mL)	Recovery (%)	CV (%)	*n*	Measured (ng/mL)	Recovery (%)	CV (%)
2500	3	2165.54 ± 18.03	86.62 ± 0.72	0.83	3	2190.13 ± 42.49	87.6 ± 1.7	1.94
600	3	578.74 ± 36.56	96.46 ± 6.09	6.31	3	597.1 ± 24.89	99.5 ± 4.15	4.17
150	3	143.55 ± 6.39	95.7 ± 4.26	4.45	3	152.71 ± 11.39	101.8 ± 7.59	7.46
Average	-	-	92.93 ± 3.69	3.87	-	-	96.31 ± 4.48	4.52

Table 2. The detection result of αB-CTX toxin in real samples.

Samples		OD450 nm Value	Detection Results
Oncomelania hupensis Gredler		0.9983 ± 0.0366	-
Shellfish (qīng é)		1.001 ± 0.0669	-
Ruditapes philippinarum		0.953 ± 0.0321	-
Viviparidae		1.0143 ± 0.0598	-
Thais clavigera Kuster		0.968 ± 0.0363	-
αB-CTX		0.0545 ± 0.0048	+
PBS		0.9805 ± 0.1158	-

Detection results of αB-CTX in real samples (*n* = 3). +: indicates that αB-CTX is detected in real samples. -: indicates that no αB-CTX is detected in real samples.

3. Discussion

The previous study in our laboratory has shown that the monomer αB-CTX cannot induce the production of an antiserum with high binding activity, failing to stimulate a strong immune effect (data not shown). When smaller antigens, especially the ploy-peptides with low molecular weights, were used, the immunogenicity was often limited or invalid [5]. To effectively solve this critical problem, the four original genes of αB-CTX were connected in series to form tetramers in this study, and the purified MBP-LK-αB-CTX4 fusion protein was used as an immunogen to elicit a strong response for antibody production. Compared to the control, the titer of serum from mice immunized by MBP-LK-αB-CTX4 was very high. In theory, the polymerization state of the antigen is an important factor affecting the immunogenicity, and the immunogenicity of the polymer is usually stronger than that of the monomer. The fusion protein MBP-LK-αB-CTX4 containing tetramer form has a larger molecular weight and better protein folding, which is easier to stimulate the immune effect of the body. As expected, the ELISA result from the antiserum titer also strongly confirmed this judgment. Therefore, the method used in this study by connecting antigens with a linker to form a multimer is feasible for improving the immunogenicity of the antigen, and it may provide a useful idea for solving similar critical problems in polypeptide antigens.

In the process of fusion, it should be noted that the spleen and connective tissue should be stripped clearly in a clean environment, and the optimal ratio of spleen cells and myeloma cells was controlled at 3:1~5:1. The addition of feeder layer cells is necessary for cell fusion, and the growth-promoting factors and nutrients secreted by the feeding layer of cells provide a comfortable and favorable environment for the growth of hybridomas [24]. How to maintain high activity of antibody has been a topic of constant concern. According to our experience, low temperature (0 °C~4 °C) and a solution with an accurate pH value (pH 7.2) are two necessary conditions to ensure the activity of antibody in the process of antibody purification.

To ensure the accurancy of ic-ELISA, it is necessary that the concentration of antigen and antibody should be accurately measured during the experiments by using multiple checkerboard iELISA experiments, and the dilution should be as accurate as possible. In fact, the accuracy of ic-ELISA can be affected by many factors, including the coating time, the choice of coating and block buffer, and the concentration of substrate and antibody [5]. To improve the accuracy of the ic-ELISA detection, the above factors should be maintained at the same condition throughout the system when the optimal conditions and reagents are explored clearly. The best coating time in this study is at 4 °C overnight, and the optimal dilution ratio of HPR-goat anti-mouse IgG antibody and antigen coating concentrations were 1:16,000 and 10 µg/mL, respectively.

4. Conclusions

In the present study, an effective antigen tandem design and expression format was used to increase the immunogenicity of the target antigen, further solving the key problem of antibody preparation. A hybridoma 5E4 stably generating mAb against αB-CTX was prepared for the first time through hybridoma technology. The resulting 5E4 mAb was specific to epitope 1 of αB-CTX, and the amino acids L14 and F15 were two critical sites for 5E4 recognition. The affinity of the purified 5E4 mAb was 1.02×10^8 L/mol, belonging to a high-affinity antibody. The developed ic-ELISA in this study based on 5E4 mAb has good specificity and accuracy, and the average recovery and CV were (92.93 ± 3.69)% and 3.87%, respectively. Under optimal conditions, the LOD of ic-ELISA was 81 ng/mL, with a linear range of 117–3798 ng/mL. Lastly, no αB-CTX was detected in five different actual shellfish seafood products. In conclusion, the developed ic-ELISA could be used as a useful and feasible method for the analysis and detection of αB-CTX in real samples.

5. Materials and Methods

5.1. Materials

Balb/c mice were purchased from Wushi Animal Laboratory (Shanghai, China). Hypoxanthine, aminopterin and thymidine supplement (HAT) was obtained from Sigma-Aldrich Chemical (St. Louis, MO, USA). BCA protein assay kit (CAT: PC0020) was purchased from Solarbio Life Science (Beijing, China). All other used chemicals were of analytical grade. All the animal experiments in the present study complied with the Research Ethics Committee of Fujian Agriculture and Forestry University, China (Permit No. PFMFAFU201610) [5,17].

5.2. Epitope Predication and 3D Structure Simulation

The mature peptide sequence and cysteine framework of αB-CTX were referred to as in the previous report [12], and the possible epitopes of αB-CTX were further analyzed using the B-Cell Linear Epitope Prediction tool on the Immune Epitope Database Analysis Resource server (http://tools.immuneepitope.org/bcell/) (accessed on 6 January 2022) [23,25]. Meanwhile, the 3D structure of αB-CTX was predicated on the basis of homology modeling using SWISS-MODEL (http://swiss.model.expasy.org/) (accessed on 6 January 2022) [26–28].

5.3. Protein Expression and Purification

To achieve high expression and purification of αB-CTX in *E. coli* BL21 (DE3), the DNA fragment encoding mature αB-CTX peptide was synthesized by Shanghai Biotech Co., Ltd. (Shanghai, China) after codon optimization [29]. The resulting DNA fragment was inserted into the plasmids pET32a and pBD-mbp for the construction of recombinant expression vectors, and the target proteins were expressed in *E. coli* BL21 (DE3) by IPTG induction (a final concentration of 1.0 mM) when the cultures reached an OD_{600} of 1.0 [30]. After cultivation, the fusion proteins were purified by Ni^{2+}-NTA affinity chromatography, and the protein concentration was determined by SDS-PAGE and BCA methods [18].

5.4. Animal Immunization and Titer Determination

To effectively activate the immune response of the mouse model, the purified MBP-LK-αB-CTX4 fusion protein was used as the immunogen, and the procedures of animal immunization and titer determination were performed as described with minor modifications [17,31]. Balb/C mice at the age of 7 weeks were immunized subcutaneously (100 μg MBP-LK-αB-CTX4 emulsified in Freund's complete adjuvant) with an interval of 2 weeks. A total of 5 days after the 5th immunization, the ELISA was used to test the titer of serum from the control and each immunized mouse. The mouse with the highest antibody titer was regarded as the spleen donor for cell fusion and hybridoma screening.

5.5. Screening and Identification of Positive Hybridoma

The detailed protocols of cell fusion and hybridoma selection were carried out as described in the referred methods [5,12]. The subclass of the antibody secreted by the positive hybridoma was identified by using Mouse Monoclonal Antibody Subtyping Kit [21], and the chromosome numbers of selected hybridoma cells were observed and counted using Jimsey staining solution as described [5].

5.6. Purification and Identification of mAb against αB-CTX

To obtain a monoclonal antibody (mAb) with high binding activity and specificity for further investigation, healthy Balb/C mice aged 8–12 weeks were sensitized by paraffin at the abdominal cavity, and then the cultured positive hybridoma (about 10^6 cells) was injected into the abdominal cavity of the mice for ascites production. The caprylic/ammonium sulfate precipitation and protein A/G methods were used to purify the IgG mAb against αB-CTX from ascites fluid, and the concentration and fineness of purified antibody were analyzed by the SDS-PAGE and BCA kits [5]. The binding activity (affinity) and specificity of the purified anti-αB-CTX mAb were determined by indirect ELISA as described [19].

5.7. Analysis of Different Binding Epitopes

To investigate the binding epitope of the prepared mAb to αB-CTX, αB-CTX was divided into three different fragments according to the protein sequence, and the resulting fragments were designated as αB-CTX (1–20), αB-CTX (10–30) and αB-CTX (20–40), respectively. These three polypeptides were synthesized by Shanghai Biotech Co., Ltd. (Shanghai, China) and then characterized by HPLC. Furthermore, 3 different peptides and the full-length αB-CTX protein were used as the antigen to coat the 96-well plates (10 μg/mL), and the purified 5E4 mAb in PBSM solution (1:4000 dilution) was added to the reaction wells after blocking and washing. The binding validation was then performed by ELISA as described [32].

5.8. Analysis of Specific Sites between mAb and αB-CTX Interaction

In view of the research results of antibody binding to a specific epitope, the critical residues for mAb-αB-CTX binding were further determined. A total of 8 amino acids located on epitope 1 of αB-CTX, regarded as the specific binding epitope for antigen-antibody interactions, were individually mutated to alanine [32]. In the same way, the constructed αB-CTX mutants were synthesized by Shanghai Biotech Co., Ltd. (Shanghai, China) and then characterized by HPLC. Then, 8 mutated αB-CTX proteins and αB-CTX were used as the antigen to coat the 96-well plates (10 μg/mL), and the purified 5E4 mAb in PBSM solution (1:4000 dilution) was added to the reaction wells after blocking and washing. The binding activities of mAb to αB-CTX mutants were determined by ELISA according to the instructions described above.

5.9. Establishment of ic-ELISA and Standard Curve

To effectively detect and analyze the content of αB-CTX in samples, indirect competitive ELISA (ic-ELISA) was established via a specific antibody-antigen reaction [24]. For testing the best concentration on the calibration curve, the standard αB-CTX protein (as a complete antigen) was 2-fold serially diluted (10~1.25 μg/mL, 625~1.2125 ng/mL and 606~1.18 pg/mL, respectively) to react with the anti-αB-CTX mAb in a tube at 37 °C for 30 min. Then, the reaction mixtures were added into the reaction wells with three replicates of each concentration standard. The data from ic-ELISA was calculated and analyzed, and the typical calibration curve and linear portion of the standard curve were illustrated by using Origin 8.0 software according to the value of plotting (B/B0) against the αB-CTX concentration [24]. In ic-ELISA, B0 is the OD450 nm value of the control well without inhibitor, and B is the OD450 nm value of the well with inhibitor. B/B0 is called the inhibition rate. IC50 indicates the concentration of inhibitor when the inhibition rate reaches 50%.

5.10. αB-CTX Detection in Spiked and Real Samples by ic-ELISA

To further assess the accuracy and recovery of the developed ic-ELISA method, the PBSM (PBS containing 5% skim milk powder) solution (no αB-CTX protein) was treated with different concentrations of standard αB-CTX as spiked samples. To test and assess the recovery and accuracy of ic-ELISA detection, the inter-assay and intra-assay were performed with three replicates of each concentration standard, and the average recovery and coefficient of variation (CV) were defined as the ratio of standard deviation to the mean in recovery test [5]. At last, five different snail and shell samples purchased randomly from Fuzhou supermarkets were detected using the ic-ELISA developed in this study.

5.11. Statistical Analysis

All the experimental data were processed and analyzed by ORIGIN 8.0 (OriginLab, Northampton, MA, USA). Curve fitting and variance analysis were conducted to describe the relationship between variables. All analyses were performed with three parallel tests.

Author Contributions: Conceptualization, H.T., R.W. and S.W.; investigation, H.T., H.L., Y.G., R.C. and R.W.; methodology, M.D., S.L. and R.W.; validation, R.C., Y.G. and S.W.; writing—original draft preparation, R.W., H.T. and S.W.; writing—review and editing, R.W. and S.W.; supervision, R.W. and S.W.; project administration, R.W. and S.W. All authors have read and agreed to the published version of the manuscript.

Funding: This work was funded by the National Key Research and Development Program of China (2019YFC0312603, 2019YFC0312602), the marine economic development subsidy fund project in Fujian Province (FJHJF-L-2020-3), and the Science and Technology Innovation Project of Fujian Agriculture and Forestry University (KFA17078A).

Institutional Review Board Statement: The study was conducted according to the guidelines of the Declaration of Helsinki and approved by the Research Ethics Committee of Fujian Agriculture and Forestry University, China (Permit No. PFMFAFU201610, approval date: 26 October 2016).

Informed Consent Statement: Not applicable.

Data Availability Statement: Not applicable.

Conflicts of Interest: The authors declare that they have no known competing financial interest or personal relationships that could have appeared to influence the work reported in this paper.

References

1. Gao, B.M.; Peng, C.; Yang, J.A.; Yi, Y.H.; Hang, J.Q.; Shi, Q. Cone Snails: A Big Store of Conotoxins for Novel Drug Discovery. *Toxins* **2017**, *9*, 397. [CrossRef]
2. Kumar, P.S.; Kumar, D.S.; Umamaheswari, S. A perspective on toxicology of Conus venom peptides. Asian Pac. *J. Trop. Med.* **2015**, *8*, 337–351.
3. Prashanth, J.R.; Brust, A.; Jin, A.H.; Alewood, P.F.; Dutertre, S.; Lewis, R.J. Cone snail venomics: From novel biology to novel therapeutics. *Future Med. Chem.* **2014**, *6*, 1659–1675. [CrossRef]
4. Vetter, I.; Lewis, R.J. Therapeutic potential of cone snail venom peptides (conopeptides). *Curr. Top. Med. Chem.* **2012**, *12*, 1546–1552. [CrossRef]
5. Wang, R.Z.; Zhong, Y.F.; Wang, J.C.; Yang, H.; Yuan, J.; Wang, S.H. Development of an ic-ELISA and immunochromatographic strip based on IgG antibody for detection of ω-conotoxin MVIIA. *J. Hazard. Mater.* **2019**, *378*, 120510. [CrossRef]
6. Peng, C.; Yao, G.; Gao, B.M.; Fan, C.X.; Bian, C.; Wang, J.; Cao, Y.; Wen, B.; Zhu, Y.; Ruan, Z.; et al. High-throughput identification of novel conotoxins from the Chinese tubular cone snail (*Conus betulinus*) by multi-transcriptome sequencing. *Gigascience* **2016**, *5*, 17. [CrossRef]
7. Barghi, N.; Concepcion, G.P.; Olivera, B.M.; Lluisma, A.O. High conopeptide diversity in Conus tribblei revealed through analysis of venom duct transcriptome using two high-throughput sequencing platforms. *Mar. Biotechnol.* **2015**, *17*, 81–98. [CrossRef]
8. Dutertre, S.; Jin, A.H.; Kaas, Q.; Jones, A.; Alewood, P.F.; Lewis, R.J. Deep venomics reveals the mechanism for expanded peptide diversity in cone snail venom. *Mol. Cell. Proteom.* **2013**, *12*, 312–329. [CrossRef]
9. Robinson, S.D.; Norton, R.S. Conotoxin gene superfamilies. *Mar. Drugs* **2014**, *12*, 6058–6101. [CrossRef]
10. Eisapoor, S.S.; Jamili, S.; Shahbazzadeh, D.; Ghavam Mostafavi, P.; Pooshang Bagheri, K. A New, High Yield, Rapid, and Cost-Effective Protocol to Deprotection of Cysteine-Rich Conopeptide, Omega-Conotoxin MVIIA. *Chem. Biol. Drug Des.* **2016**, *87*, 687–693. [CrossRef]
11. Rigo, F.K.; Dalmolin, G.D.; Trevisan, G.; Tonello, R.; Silva, M.A.; Rossato, M.F.; Klafke, J.Z.; Cordeiro Mdo, N.; Castro Junior, C.J.; Montijo, D.; et al. Effect of omega-conotoxin MVIIA and Phalpha1beta on paclitaxel-induced acute and chronic pain. *Pharmacol. Biochem. Behav.* **2013**, *114*, 16–22. [CrossRef] [PubMed]
12. Luo, S.; Christensen, S.; Zhangsun, D.; Wu, Y.; Hu, Y. A Novel Inhibitor of a9a10 Nicotinic Acetylcholine Receptors from Conus vexillum Delineates a New Conotoxin Superfamily. *PLoS ONE* **2013**, *8*, e54648.
13. Albuquerque, E.X.; Pereira, E.F.; Alkondon, M.; Rogers, S.W. Mammalian nicotinic acetylcholine receptors: From structure to function. *Physiol. Rev.* **2009**, *89*, 73–120. [CrossRef]
14. Holtman, J.R.; Dwoskin, L.P.; Dowell, C.; Wala, E.P.; Zhang, Z. The novel small molecule alpha9alpha10 nicotinic acetylcholine receptor antagonist ZZ204G is analgesic. *Eur. J. Pharmacol.* **2011**, *670*, 500–508. [CrossRef]
15. Zheng, G.; Zhang, Z.; Dowell, C.; Wala, E.; Dwoskin, L.P. Discovery of non-peptide, small molecule antagonists of alpha9alpha10 nicotinic acetylcholine receptors as novel analgesics for the treatment of neuropathic and tonic inflammatory pain. *Bioorg. Med. Chem. Lett.* **2011**, *21*, 2476–2479. [CrossRef]
16. Chernyavsky, A.I.; Arredondo, J.; Vetter, D.E.; Grando, S.A. Central role of alpha9 acetylcholine receptor in coordinating keratinocyte adhesion and motility at the initiation of epithelialization. *Exp. Cell. Res.* **2007**, *313*, 3542–3555. [CrossRef]
17. Wang, R.Z.; Wang, J.C.; Liu, H.M.; Gao, Y.H.; Zhao, Q.; Ling, S.M.; Wang, S.H. Sensitive immunoassays based on specific monoclonal IgG for determination of bovine lactoferrin in cow milk samples. *Food Chem.* **2021**, *338*, 127820. [CrossRef]
18. Wang, R.Z.; Gu, X.S.; Zhuang, Z.H.; Zhong, Y.F.; Yang, H.; Wang, S.H. Screening and molecular evolution of a single chain variable fragment antibody (scFv) against Citreoviridin toxin. *J. Agric. Food Chem.* **2016**, *64*, 7640–7648. [CrossRef]
19. Wang, R.Z.; Zeng, L.M.; Yang, H.; Zhong, Y.F.; Wang, J.C.; Ling, S.M.; Saeed, A.F.; Yuan, J.; Wang, S.H. Detection of okadaic acid (OA) using ELISA and colloidal gold immunoassay based on monoclonal antibody. *J. Hazard. Mater.* **2017**, *339*, 154–160. [CrossRef]
20. Cai, P.Y.; Wang, R.Z.; Ling, S.M.; Wang, S.H. A high sensitive platinum-modified colloidal gold immunoassay for tenuazonic acid detection based on monoclonal IgG. *Food Chem.* **2021**, *360*, 130021. [CrossRef]
21. Ling, S.M.; Zhao, Q.; Iqbal, M.N.; Dong, M.K.; Li, X.L.; Lin, M.; Wang, R.Z.; Lei, F.Y.; He, C.Z.; Wang, S.H. Development of immunoassay methods based on monoclonal antibody and its application in the determination of cadmium ion. *J. Hazard. Mater.* **2021**, *411*, 124992. [CrossRef]
22. Fadlalla, M.H.; Ling, S.M.; Wang, R.Z.; Li, X.L.; Yuan, J.; Xiao, S.W.; Wang, K.; Tang, S.Q.; Elsir, H.; Wang, S.H. Development of ELISA and lateral flow immunoassays for ochratoxins (OTA and OTB) detection based on monoclonal antibody. *Front. Cell. Infect. Microbiol.* **2019**, *10*, 80. [CrossRef] [PubMed]
23. Liew, O.W.; Ling, S.S.M.; Lilyanna, S.; Zhou, Y.; Wang, P.P.; Chong, J.P.C.; Ng, Y.X.; Lim, A.E.S.; Leong, E.R.Y.; Lin, Q.F.; et al. Epitope-directed monoclonal antibody production using a mixed antigen cocktail facilitates antibody characterization and validation. *Commun. Biol.* **2021**, *4*, 441. [CrossRef] [PubMed]
24. Ling, S.M.; Chen, Q.A.; Zhang, Y.M.; Wang, R.Z.; Jin, N.; Pang, J.; Wang, S.H. Development of ELISA and colloidal gold immunoassay for tetrodotoxin detection based on monoclonal antibody. *Biosens. Bioelectron.* **2015**, *71*, 256–260. [CrossRef] [PubMed]
25. Larsen, J.E.; Lund, O.; Nielsen, M. Improved method for predicting linear B-cell epitopes. *Immunome Res.* **2006**, *2*, 2. [CrossRef] [PubMed]

26. Biasini, M. SWISS-MODEL: Modelling protein tertiary and quaternary structure using evolutionary information. *Nucleic Acids Res.* **2014**, *42*, 252–258. [CrossRef]
27. Arnold, K.; Bordoli, L.; Kopp, J.; Schwede, T. The SWISS-MODEL Workspace: A web-based environment for protein structure homology modelling. *Bioinformatics* **2006**, *22*, 195–201. [CrossRef]
28. Guex, N.; Peitsch, M.C.; Schwede, T. Automated comparative protein structure modeling with SWISS-MODEL and Swiss-PdbViewer: A historical perspective. *Electrophoresis* **2009**, *30*, S162–S173. [CrossRef]
29. Jia, K.Z.; Zhang, D.P.; Wang, Y.H.; Liu, Y.J.; Kong, X.Z.; Yang, Q.H.; Chen, H.L.; Xie, C.J.; Wang, S.H. Generation and characterization of a monoclonal antibody against human BCL6 for immunohistochemical diagnosis. *PLoS ONE* **2019**, *14*, e0216470. [CrossRef]
30. Yang, H.; Zhong, Y.F.; Wang, J.C.; Zhang, Q.H.; Li, X.L.; Ling, S.M.; Wang, S.H.; Wang, R.Z. Screening of a scFv antibody with high affinity for application in Human IFN-γ immunoassay. *Front. Microbiol.* **2018**, *9*, 261. [CrossRef]
31. Ling, S.M.; Li, X.L.; Zhao, Q.; Wang, R.Z.; Tan, T.; Wang, S.H. Preparation of monoclonal antibody against penicillic acid (PA) and its application in the immunological detection. *Food Chem.* **2020**, *319*, 126505. [CrossRef] [PubMed]
32. Wang, R.Z.; Xiang, S.S.; Zhang, Y.F.; Chen, Q.Y.; Zhong, Y.F.; Wang, S.H. Development of a functional antibody by using a green fluorescent protein frame as the Template. *Appl. Environ. Microb.* **2014**, *80*, 4126–4137. [CrossRef] [PubMed]

Article

Zearalenone Affect the Intestinal Villi Associated with the Distribution and the Expression of Ghrelin and Proliferating Cell Nuclear Antigen in Weaned Gilts

Quanwei Zhang [1,†], Libo Huang [1,†], Bo Leng [1], Yang Li [1], Ning Jiao [1], Shuzhen Jiang [1], Weiren Yang [1,*] and Xuejun Yuan [2,*]

1 College of Animal Sciences and Technology, Shandong Agricultural University, Tai'an City 271018, China; sdndzqw@163.com (Q.Z.); huanglibo@sdau.edu.cn (L.H.); lengbo666@163.com (B.L.); liyang_cc@yeah.net (Y.L.); jiaoning@sdau.edu.cn (N.J.); shuzhen305@163.com (S.J.)
2 College of Life Sciences, Shandong Agricultural University, Tai'an City 271018, China
* Correspondence: wryang@sdau.edu.cn (W.Y.); xjyuan@sdau.edu.cn (X.Y.); Tel.: +86-186-0548-9796 (W.Y.); +86-134-7538-6175 (X.Y.)
† Co-first author.

Abstract: This study explored and investigated how zearalenone (ZEA) affects the morphology of small intestine and the distribution and expression of ghrelin and proliferating cell nuclear antigen (PCNA) in the small intestine of weaned gilts. A total of 20 weaned gilts (42-day-old, D × L × Y, weighing 12.84 ± 0.26 kg) were divided into the control and ZEA groups (ZEA at 1.04 mg/kg in diet) in a 35-d study. Histological observations of the small intestines revealed that villus injuries of the duodenum, jejunum and ileum, such as atrophy, retardation and branching dysfunction, were observed in the ZEA treatment. The villi branch of the ileum in the ZEA group was obviously decreased compared to that of the ileum, jejunum and duodenum, and the number of lymphoid nodules of the ileum was increased. Additionally, the effect of ZEA (1.04 mg/kg) was decreased by the immunoreactivity and distribution of ghrelin and PCNA in the duodenal and jejunal mucosal epithelial cells. Interestingly, ZEA increased the immunoreactivity of ghrelin in the ileal mucosal epithelial cells and decreased the immunoreactivity expression of PCNA in the gland epithelium of the small intestine. In conclusion, ZEA (1.04 mg/kg) had adverse effects on the development and the absorptive capacity of the villi of the intestines; yet, the small intestine could resist or ameliorate the adverse effects of ZEA by changing the autocrine of ghrelin in intestinal epithelial cells.

Keywords: zearalenone; weaned gilt; intestinal morphology; ghrelin; PCNA

Key Contribution: The results showed that ZEA (1.04 mg/kg) in a diet could damage the intestinal structure and significantly affect the expression of PCNA and ghrelin in weaned gilt.

Citation: Zhang, Q.; Huang, L.; Leng, B.; Li, Y.; Jiao, N.; Jiang, S.; Yang, W.; Yuan, X. Zearalenone Affect the Intestinal Villi Associated with the Distribution and the Expression of Ghrelin and Proliferating Cell Nuclear Antigen in Weaned Gilts. *Toxins* **2021**, *13*, 736. https://doi.org/10.3390/toxins13100736

Received: 6 September 2021
Accepted: 15 October 2021
Published: 19 October 2021

1. Introduction

Fusarium, a kind of fungi, is widely distributed in nature and is common in North America, Asia and Europe with mild climates [1]. Mycotoxins are produced by *Fusarium*, which is a major and serious threat to animal and human health, as well as livestock production [2–4]. In terms of animal health and productivity, the most important mycotoxins were trichothecenes, zearalenone (ZEA), deoxynivalenol and fumonisins B_1 [5,6]. The gastrointestinal tract is one of the most sensitive tissues to these mycotoxins [1]. Studies have shown that mycotoxins can damage animal intestines through impairing the reduction–oxidation reaction balance of the body, affecting the digestive tract function and causing intestinal villus atrophy and an inflammatory response in the intestinal epithelial cells of piglets [3,7,8]. Some in vitro studies have shown that food contaminated by ZEA and ZEA metabolites can affect the synthesis of porcine cytokines and the structural integrity of the intestinal epithelium [9,10]. The results of the differential gene expression and microarrays showed that there were 190 differently expressed genes in isolated IPEC-1 (porcine epithelial cells) treated with ZEA, of which 70% were upregulated [11]. A low dose (40-μg/kg BW) of ZEA did not change the mucosa thickness, villus length and villus-to-crypt of the duodenum [12]. As far as we know, there are few published in vivo studies available on the evaluation of the impacts of ZEA exposure on the intestinal structure and function of gilts.

The small intestine is a critical place where nutrients are absorbed. Intestinal epithelial cells are the first important and physical barrier to avoid the gastrointestinal absorption of toxins [13], as well as the first target of toxins [14,15]. Ghrelin is a pleiotropic hormone, which can promote growth hormone secretion [16], increase food intake [17], regulate the energy metabolism and intracellular homeostasis [18,19] and is even involved in the immune regulation of the intestinal mucosa [20]. Blood ghrelin mainly comes from the gastrointestinal tract; a small amount of circulating ghrelin comes from the immune system, kidney, pancreas, testis, ovary and placenta [21]. Research in pigs showed that ghrelin also influenced or regulated their growth and development [22]. The results from Willemen illustrated that the amount of expressing active ghrelin in gastric cells in the normal weight neonates was higher compared with the small-for-gestational age piglets [23]. A previous study reported that a diet containing ZEA could affect the ovarian histology and follicular development by affecting the expression of PCNA and ghrelin in the ovaries [24] and increasing the PCNA expression of granulosa cells and then accelerate the changes in ovarian histology and the development of ovaries in the weaned gilts [24]. However, information was limited on the effects of a diet containing ZEA exposure on the expression and distribution of ghrelin and PCNA in the small intestine of weaned gilts.

The purpose of this research was to evaluate the impacts of ZEA in the diet on the histological structure and the mRNA and protein expressions of ghrelin and PCNA in the small intestine of weaned gilts. The results will be helpful to evaluate the impacts of a 1.04-mg/kg dose of ZEA in a diet on the subsequent damage and the digestive capacity of the small intestines in weaned gilts.

2. Results

2.1. Serum ZEA, α-ZOL, and β-ZOL

The results showed that the serum α-ZOL (α-zearalenol), β-ZOL (β-zearalenol) and ZEA contents in the ZEA group were higher than those in the control group (Table 1, $p < 0.05$). These confirmed that ZEA could be absorbed and partially degraded into β-ZOL and α-ZOL by the liver.

Table 1. Zearalenone contents in the serum of weaned gilts (μg/mL).

Items	Zearalenone	α-Zearalenol	β-Zearalenol
Control	0.00 ± 0.00 [b]	0.291 ± 0.02 [b]	0.035 ± 0.01 [b]
ZEA1.0	0.15 ± 0.04 [a]	0.91 ± 0.10 [a]	0.77 ± 0.01 [a]

Data were presented by the mean ± SD ($n = 6$). [a,b] Means in the same column were significantly different ($p < 0.05$).

2.2. Morphological Structure and Measurement

Compared with the gilts in the control treatment, the gilts in the ZEA treatment showed intestinal villus injuries, such as shortening, retardation and branching dysfunction, and more obvious morphological changes of the villus branch in the jejunum and ileum (Figure 1B1–C4). The duodenal villi in the ZEA treatment displayed a finger-like structure (Figure 1A3,A4), but the duodenal villus length of the gilts in the control treatment was longer than those of the ZEA treatment (Figure 1A1–A4). The jejunal villi of the gilts in ZEA treatment group were stubby and leaf-like (Figure 1B3,B4), but the jejunal villi of the gilts in the control treatment were finger-like (Figure 1B1,B2). The branches of the ileal villi in the ZEA treatment were obviously incomplete (Figure 1C3,C4), and the number and the area of the ileal submucosal lymph nodes were increased.

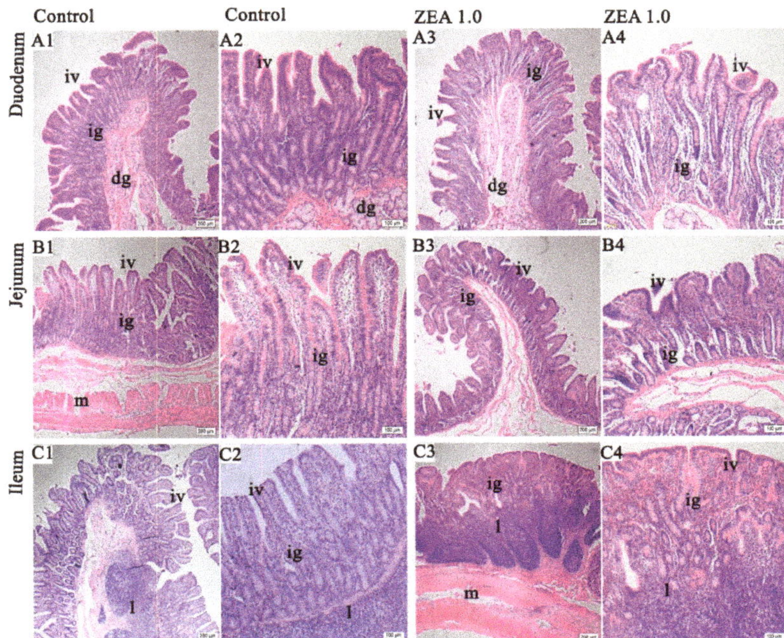

Figure 1. Representative hematoxylin and eosin staining images of the small intestine in weaned gilts (*n* = 6). (**A1,A2,B1,B2,C1,C2**) were the control treatment, and (**A3,A4,B3,B4,C3,C4**) were the ZEA1.0 treatment. Scale bars were 200 μm for (**A1,A3,B1,B3,C1,C3**) and 100 μm for (**A2,A4,B2,B4,C2,C4**), respectively. ig represented intestinal gland, iv represented intestinal villus, g represented duodenal gland, l represented lymphoid nodule and m represented musculari.

The parameters of the morphometric analysis of the small intestine in the control and ZEA treatment are listed in Table 2. The results showed that the villi lengths-to-crypt depth and villus length of ZEA exposure were significantly reduced ($p < 0.05$), and the crypt depth of the ZEA treatment became thick ($p < 0.05$) compared to the control treatment in the intestinal three segments.

Table 2. Morphometric analysis of the small intestine in weaned gilts.

Items		Villus Length (μm)	Crypt Depth (μm)	VL/CD
Duodenum	Control	515.91 ± 7.43 [a]	713.31 ± 7.58 [b]	0.73 ± 0.01 [a]
	ZEA1.0	355.86 ± 7.66 [b]	880.36 ± 12.76 [a]	0.41 ± 0.01 [b]
Jejunum	Control	1059.37 ± 19.13 [a]	493.21 ± 7.03 [b]	2.16 ± 0.05 [a]
	ZEA1.0	376.12 ± 13.49 [b]	847.03±7.01 [a]	0.59 ± 0.03 [b]
Ileum	Control	334.67 ± 9.85 [a]	895.51 ± 10.68 [b]	0.48 ± 0.01 [a]
	ZEA1.0	178.83 ± 8.68 [b]	1759.83 ± 23.93 [a]	0.10 ± 0.01 [b]

Data were presented by the mean ± SD (*n* = 6). [a,b] Means in the same column were significantly different ($p < 0.05$). VL/CD, villus length/crypt depth.

2.3. The Ghrelin Immunoreactive Cells Distribution

The distribution and expression of ghrelin in the small intestine were presented in Figure 2. The immunohistochemical results showed that ghrelin-positive substances were distributed in the cytoplasm of the villi and glandular epithelium. The ghrelin immunoreactivity was stronger in the villi mucosal epithelium adjacent to the intestinal lumen at the apical surface of the folds, but it was gradually reduced and weakened in epithelial cells on both sides and near the base. The ghrelin localization pattern of the duodenum (Figure 2A1–B3) and jejunum (Figure 2C1–D3) in the control treatment (Figure 2A1–A3,C1–C3) was essentially the same as that in the ZEA treatment (Figure 2B1–B3,D1–D3), but in the ileal mucosal epithelial cells, ZEA increased the ghrelin immunoreactivity compared to the control (Figure 2E1–F3). The results of a single villus' integrated optic density (SIOD) of duodenal and jejunal ghrelin in weaned gilts were consistent with the above results of the immunochemical analysis. The SIOD of the ZEA group were obviously decreased compared to those in the control treatment ($p < 0.05$, Figure 3B). However, the IOD and SIOD of the ileal ghrelin in the ZEA group were increased significantly compared to those in the control group (Figure 3A,B, $p < 0.05$).

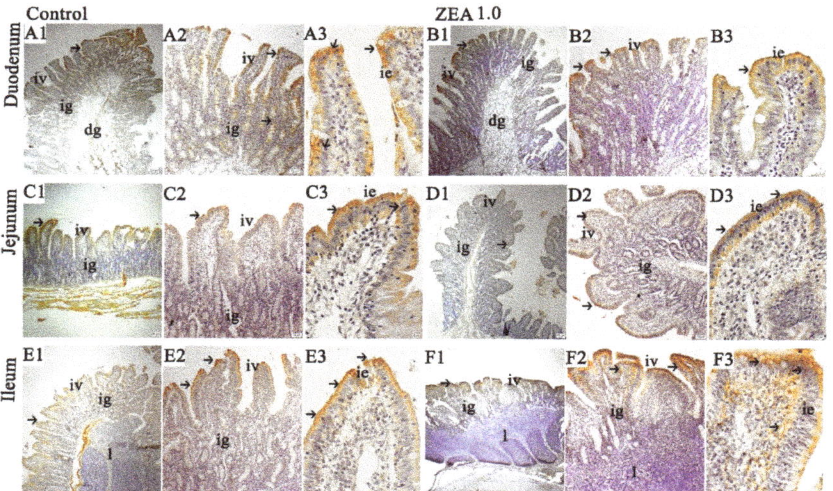

Figure 2. Representative distribution of ghrelin immuno-positive cells in the small intestine of weaned gilts ($n = 6$). (**A1–A3,C1–C3,E1–E3**) were the control, and (**B1–B3,D1–D3,F1–F3**) were the ZEA1.0 treatment. Scale bars of (**A1,B1,C1,D1,E1,F1**) were 200 μm, of (**A2,B2,C2,D2,E2,F2**) were 100 μm and of (**A3,B3,C3,D3,E3,F3**) were 20 μm. ig represented the intestinal glands, iv represented the intestinal villus, dg represented the duodenal gland, ie represented the intestinal villus epithelium, l represented the lymphoid nodule and m represented the muscularis.

2.4. The Distribution of PCNA Immunoreactive Cells

The expression and distribution of PCNA in the small intestines of gilts were detected (Figure 4). The immunohistochemical results showed that PCNA-positive substances were distributed in small intestinal villi and intestinal glands. The PCNA immunoreactivity was stronger in villus epithelial cells, especially at the bottom of the small intestinal villi, but weakened gradually in the villus epithelial cells of both sides and at the top of the small intestinal villi. The location pattern of PCNA in the duodenum (Figure 4A1–A4), jejunum (Figure 4C1–C4) and ileum (Figure 4E1–E4) of the control treatment were basically the same as those in the ZEA treatment (Figure 4B1–B4,D1–D4,F1–F4). The results of the ZEA treatment showed that the PCNA immunoreactivity of the intestinal gland cells in the duodenum, jejunum, ileum (Figure 4A4,B4,C4,D4,E4,F4) and villus epithelial cells at the base of the small intestinal villi (Figure 4A3,B3,C3,D3,E3,F3) were significantly weaker than those of the control treatment. The results of the IOD of duodenal and jejunal PCNA and SIOD in the duodenal, jejunal and ileal PCNA revealed that those of the ZEA group were significantly lower than those of the control group (Figure 3C,D, $p < 0.05$).

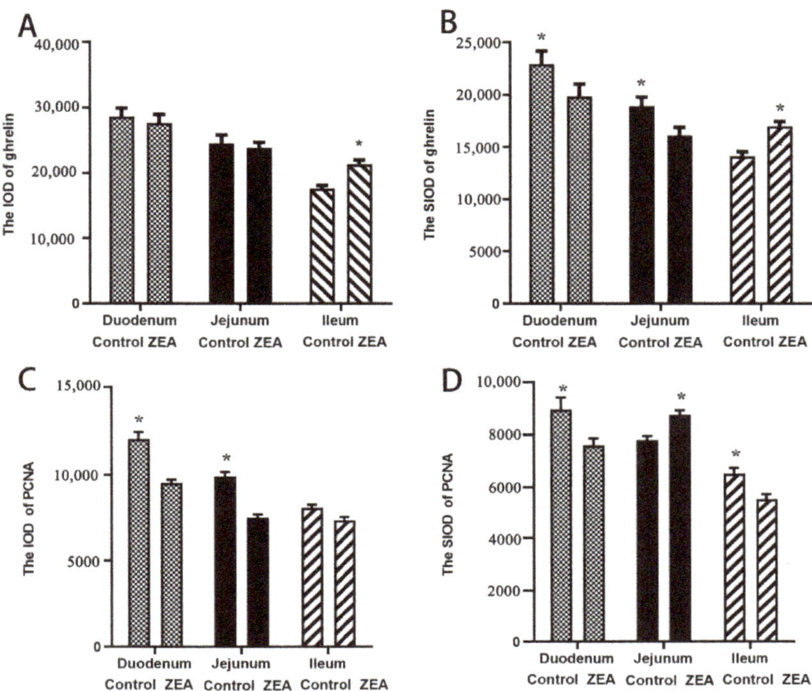

Figure 3. The IOD and SIOD of ghrelin and PCNA of the small intestine in weaned gilts ($n = 6$). * Means were significantly different ($p < 0.05$). (**A**) The IOD of ghrelin, (**B**) the SIOD of ghrelin, (**C**) the IOD of PCNA and (**D**) the SIOD of PCNA.

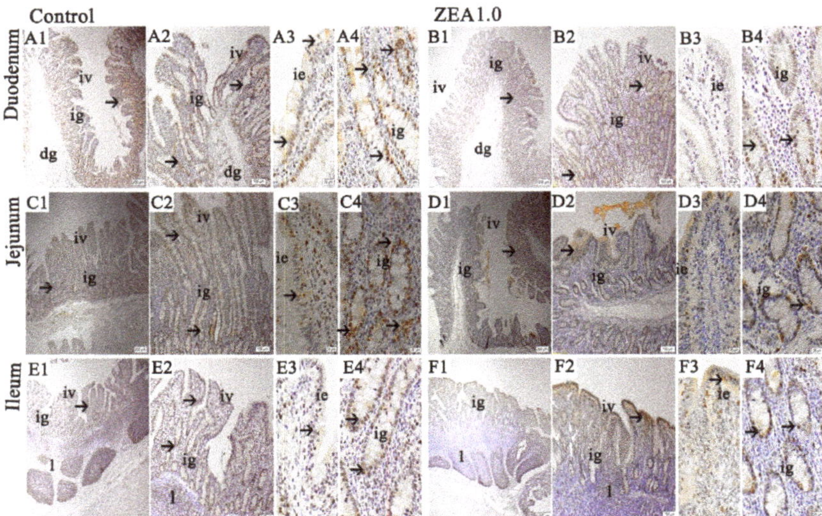

Figure 4. Representative distribution of PCNA immuno-positive cells of the small intestine in weaned gilts. (**A1–A4,C1–C4,E1–E4**) were the control treatment, and (**B1–B4,D1–D4,F1–F4**) were the ZEA1.0 treatment. Scale bars were 200 μm for (**A1,B1,C1,D1,E1,F1**), 100 μm for (**A2,B2,C2,D2,E2,F2**) and 20 μm for (**A3,A4,B3,B4,C3,C4,D3,D4,E3,E4,F3,F4**). ig represented the intestinal gland, iv represented the intestinal villus, dg represented the duodenal gland, ie represented the intestinal villus epithelium, l represented the lymphoid nodule and m represented the muscularis.

2.5. The mRNA and Protein Relative Expressions of Ghrelin and PCNA

The mRNA relative expression of ghrelin was consistent with the immunohistochemistry analysis above. Compared to the control treatment, the decreased mRNA relative expression of the ileal and jejunal PCNA (Figure 5A, $p < 0.05$) were also observed in the ZEA treatment, but there were no obvious differences in the duodenal ghrelin and PCNA mRNA relative expressions that were observed ($p > 0.05$) between the control and ZEA treatments (Figure 5A).

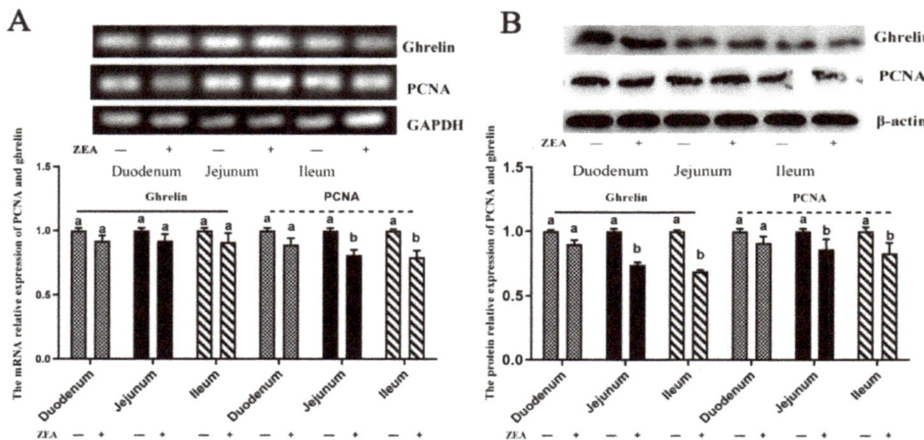

Figure 5. The protein and mRNA relative expressions of ghrelin and PCNA in the small intestine (duodenum, jejunum and ileum) in weaned gilts ($n = 4$). (**A**) The mRNA relative expressions of ghrelin and PCNA, and (**B**) the protein expressions of ghrelin and PCNA. [a,b] Means differ significantly ($p < 0.05$).

The results from the protein expressions of ghrelin and PCNA in the jejunum and ileum showed that the ZEA-treated gilts were significantly decreased compared to the control gilts ($p < 0.05$, Figure 5B).

3. Discussion

The meaningful findings of this study were that ZEA might damage the intestinal structure by changing the expression of ghrelin and PCNA. It was found that prepubertal pigs might be very sensitive animals to ZEA toxicity [25]. The estrogenic properties of ZEA have been reported extensively in the literature. Yang et al. [26] and Zinedine et al. [27] found that ZEA caused a high estrogen syndrome in animals, leading to reproductive disorders and infertility, ovarian and uterine dilation and decreased pregnancy rates in pigs and cattle [26,27]. Importantly, the small intestine acts as the first line of defense for ZEA, which is mainly absorbed into the intestinal tract and can cause intestinal damage. Liu et al. [28] reported that HSP70 expression and MDA content in the small intestine (duodenum, jejunum and ileum) were increased in weaned gilts fed a 1.04-mg/kg ZEA diet. The meaningful findings of this study were that ZEA might damage the intestinal structure by changing the expression of ghrelin and PCNA.

3.1. Morphological Structure of Small Intestine and Serum ZEA, α-ZOL and β-ZOL

Studies have shown that the small intestine is the key and main part of the absorption of most nutrients. Therefore, the changes in the structure and function of the small intestinal mucosa are closely related to the digestion and absorption of nutrients. Moreover, the villi, finger-like protrusions of the small intestine expand the surface area of the mucosa and are arranged within the intestinal mucosa epithelial cell layer facing the lumen to form a protective barrier to protect the body from direct contact with microorganisms and food antigens [29]. ZEA can cause intestinal damage, which would lead to the impairment of the intestinal absorption capacity and barrier function [30,31]. Intestinal morphology changes caused by mycotoxins (deoxynivalenol and ZEA) affected the defense mechanisms of the large intestine, and the number of plasma cells and lymphocytes also increased [8,32–35]. The report confirmed that immature gilts were administered ZEA (40-µg/kg BW) orally one week, and the number of lymphocytes and goblet cells in the intestinal villus mucosal epithelium obviously increased [36–38]. Especially the results of the ileum and jejunum in this study, ZEA exposure damaged the structure of the small intestine by shortening the villi and destroying the branching function of the ileum and jejunum and the decreased villus length of the duodenum, jejunum and ileum treated with ZEA. In the current study, it was hinted that 1.04-mg/kg ZEA resulted in the reduction of the functional mucosal epithelial surface area; the nutrient absorption capacity might decrease in a short time period. However, Liu et al. [28] found no significant difference in the ADFI, ADG or feed efficiency (ADG/ADFI) in gilts when the gilts were treated with ZEA at 1.04 mg/kg in a diet) for 35 days. Therefore, the grow performance index was not a sensitive index of short-term intestinal injury, the morphological structure and the index of the length of the small intestine villi, and the depth of the crypts was more accurate for evaluating the effect of ZEA on the intestinal morphology. This result also prompted that ZEA was mainly absorbed through small intestinal cells; after enterohepatic circulation, ZEA was degraded into β-ZOL and α-ZOL by the liver and then combined with glucuronic acid. Although ZEA and its metabolites were finally excreted through feces, urine or milk, ZEA and its metabolites would still accumulate in target organs in animals and endanger animal health [39]. The higher contents of α-ZOL, β-ZOL and ZEA in the serum of the ZEA treatment indicated that dietary 1.04-mg/kg ZEA could be absorbed by intestinal epithelial cells and degraded in the liver and act on reproductive organs and other target organs through blood circulation.

3.2. The Distribution and Expression of Ghrelin

Ghrelin, a brain–gut peptide consisting of 26 amino acids, has attracted widespread attention since it was discovered and plays a protective role in animal gastrointestinal injury [40]. Ghrelin is most densely distributed in the gastric and intestine mucosa of various vertebrates (mammals and nonmammals) [41,42]. The intestinal recess and villous cells greatly reduce the number of ghrelin with the backward extension of the digestive tract [43]. Reports were confirmed that the main secretory segment of ghrelin was the duodenum [44]; the number of ghrelin results in this research showed that the secretion of ghrelin in the duodenal and jejunal mucosa of the ZEA treatment decreased, which might be the normal physiological defense response of intestinal epithelial cells to a reduced ZEA absorption. The function of the jejunum is to digest and absorb food. Moreover, ghrelin was found to affect the food intake, endocrine regulation of intestinal emptying and motility in rats/mice [45], and in human metabolic activities, it has been reported that ghrelin is associated with intestinal mucosa [21,46,47]. The domestic research showed that ghrelin acted as a gastrointestinal hormone to stimulate the appetite, increase the feed intake and regulate the energy balance [48,49]. However, the high expression of ghrelin (IOD and SIOD) in the ileum of the control was lower than those of the ZEA treatment, which we believed that might be related the immune function of the ileum to resist the toxicity of ZEA. Our results prompted us to relate the results of the Keap1–Nrf2 signaling pathway, which was likely activated by ZEA in the ileum [50]. ZEA may have exerted influence on the hormone/signal molecule secretion rule of the small intestine. Previous studies showed that ghrelin is potentially therapeutic for mucosal injuries and intestinal permeability [51,52]. Hatoya et al. [53] found that higher estrogen receptor levels caused lower ghrelin mRNA levels. This study hinted that ZEA could cause a decrease in the secretion of ghrelin in the duodenal and jejunal mucosa, which might be the normal physiological defense response of intestinal epithelial cells to a reduced ZEA absorption.

However, its molecular mechanism needs to be further confirmed and further studied, because the effects of ghrelin on ZEA induced the toxicity of the ileum in vitro.

3.3. The Distribution and Expression of PCNA

It had been confirmed that the renewal and proliferation of intestinal villi epithelial cells were migrated from the crypt to the top of the villi and that the continuous proliferation and differentiation of intestinal stem cells in the intestinal crypts could achieve a renewal of the intestinal epithelium, so the proliferation rate of the epithelial cells and the rate of apoptosis and shedding of the epithelial cells were closely related to the growth of the intestinal villi [54]. The PCNA exists in normal proliferative cells and tumor cells, PCNA was also a major endogenous marker for testing the cell proliferation ability [55]. In addition, PCNA also plays an important role in the post-traumatic repair of many intestinal diseases. Liu et al. [56] confirmed that ZEA at 1.04 mg/kg in a diet exerted immunotoxicity and cytotoxicity through inducing oxidative stress, and then led to apoptosis and DNA oxidative damage. However, there were reports that an alternative pathway of DNA damage was closely connected to the monoubiquitination of PCNA [57,58]. In the current study, we also found that ZEA decreased the expression of PCNA in the villus epithelium and increased the expression of PCNA in the intestinal gland, which was consistent with ZEA destroying the branching and damaging the structure of the small intestine. The increased expression of PCNA in small intestinal glands showed the importance of PCNA, which was related to the self-repair mechanism of the small intestine. In addition, the results of the PCNA in the present study also indicated the influence and damage of 1.04-mg/kg ZEA to the gilts' small intestines, but their bodies still had the ability to be repaired; however, we suspected that the repair ability of the small intestine in the resistance to ZEA injury and the effect of ZEA on PCNA expression might occur via destroying the proliferation ability of small intestinal stem cells, which drive the renewal and rebirth of the intestinal epithelium every 2 to 3 days [59]. Therefore, further studies were needed to study the damaging effects of ZEA on intestinal stem cells and clarify the potential toxic effect of ZEA.

4. Conclusions

In conclusion, under the experimental conditions, we detected the influence of 1.04-mg/kg ZEA on the morphological structure of the small intestine in weaned gilts and the expression and distribution of ghrelin and PCNA, which act as important markers of intestinal development and functionality. The results also suggested that ZEA might have adverse effects on the health and growth of gilts at a later stage. More in vivo and in vitro studies are needed to confirm the relationship among the nutrient absorption, growth performance of pigs and intestinal injury caused by ZEA.

5. Materials and Methods

5.1. Ethics Statement, Experimental Design, Animals and Treatments

In this experiment, the gilts were fed in accordance with the Guide for the Care and Use of Laboratory Animals, which were approved by the Committee on the Ethics of Shandong Agricultural University (ID: S20180058, date of approval 10 December 2019).

Twenty healthy D $\times$ L $\times$ Y weaned gilts (Duroc $\times$ Landrace $\times$ Yorkshire) aged 42 ± 2 d (average weight 12.84 ± 0.26 kg) were placed in individual 0.48 m^2 cages. All gilts were fed ad libitum and had free access to water, the relative humidity of the room was approximately 65% and the temperature maintained between 26 and 28 °C during the whole experimental period. Gilts were randomly divided into two treatments (10 gilts per treatment) to receive a basal diet or a 1.0-mg/kg dose ZEA diet (basal diet added with 1.0-mg/kg ZEA). The zearalenone dosage used in this current study was due to previous investigations and the recent literature [60–63], according to Liu et al. [28], for 35 d following a 7-d adaptation.

The basal diet (Table 3) was selected to meet or exceed the piglets' nutrient requirements recommended by the NRC 2012 (National Research Council). The ZEA diet was prepared according to the studies by Dai et al. [24] and Liu et al. [28]. The two diets were produced in the same batch and stored in prior to feeding. The nutrient composition of the experimental diets was analyzed according to the method in which were described by the 2012 AOAC (Association of Official Analytical Chemists). According to Zhou et al. [64] and Liu et al. [28], the diet sample was collected before and at the end of the experiment; then, we determined the toxin contents in the diets (Table 4) The LC-MS method high-performance liquid chromatography (HPLC) tandem mass spectrometry (MS) with the fluorescence detection method, affinity column chromatography method and the external standard method were used to quantify FUM (fumonisin) and DON (deoxynivalenol) and AFL (aflatoxin). Using the LC method in conjunction with the fluorescence detection method, affinity column chromatography method, and the external standard method quantified ZEA. Its minimum detection concentration of the toxins was 0.05 mg/kg for DON, 1.0 µg/kg for AFL, 0.1 mg/kg for FUM and 0.01 mg/kg for ZEA.

Table 3. The compositions and ingredients of the basal diet.

Ingredients (%)	Content	Nutrients (%)	Analyzed Values
Corn	53.00	Metabolizable energy, MJ/kg	13.22
Whey powder	6.50	Crude protein	19.40
Wheat middling	5.00	L-Lysine HCl	0.30
Sodium chloride	0.20	Sulfur amino acid	0.79
Soybean oil	2.50	Total phosphorus	0.73
Limestone, Pulverized	0.30	Threonine	0.90
Fish meal	5.50	Methionine	0.46
Calcium phosphate	0.80	Tryptophan	0.25
Soybean meal	24.76	Lysine	1.36
L-threonine	0.04	DON mg/kg	0.41
DL-methionine	0.10	AFL ug/kg	1.62
Calcium	0.84	FUM mg/kg	0.28
Premix [1]	1.00	ZEA mg/kg	0.14
Total	100		

[1] Supplied per kg of diet: VE, 24 IU; VA, 3300 IU; K_3, 0.75 mg; D_3, 330 IU; B_{12}, 0.02625 mg; B_6, 2.25 mg; B_2, 5.25 mg; B_1, 1.50 mg; niacin, 22.5 mg; pantothenic acid, 15.00 mg; biotin, 0.075 mg; Mn ($MnSO_4 \cdot H_2O$), 6.00 mg; Zn ($ZnSO_4 \cdot H_2O$), 150 mg; Fe ($FeSO_4 \cdot H_2O$), 150 mg; Cu ($CuSO_4 \cdot 5H_2O$), 9.00 mg; Se (Na_2SeO_3), 0.45 mg. folic acid, 0.45 mg; I (KIO_3), 0.21 mg.

Table 4. Primers sequences of *GAPDH, ghrelin and PCNA*.

Target Genes	Accession No.	Primer Sequences	Product (bp)
Ghrelin	AF_308930	F: CCGAACACCAGAAAGTGCAG R: CGTTGAACCGGATTTCCAGC	144
PCNA	NM_001291925.1	F: GTGATTCCACCACCATGTTC R: TGAGACGAGTCCATGCTCTG	145
GAPDH	NM_001206359.1	F: ATGGTGAAGGTCGGAGTGAA R: CGTGGGTGGAATCATACTGG	154

The measured results of the toxins showed that the concentrations of the aflatoxins (DON, AFL, FUM and ZEA) in the basal diets were 0.41 mg/kg, 1.62 µg/kg, 0.28 mg/kg and 0.14 mg/kg and were 0.41 mg/kg, 1.59 µg/kg, 0.28 mg/kg and 1.04 mg/kg in the ZEA diet.

5.2. Sample Collection and Preparation

After the experiment was completed, all gilts were fasted for 10–12 h and then blood and separated serum were collected. The serum samples of the gilts were taken and kept at 20 °C for the ZEA content analysis. Collected samples (2.0–2.5 cm in length, approximately) from the middle of the duodenum, jejunum and ileum were immediately extracted after evisceration. One sample was stored in a RNase-free frozen tube, then placed and immersed in the liquid nitrogen immediately and kept at the ultra-low temperature of −80 °C in a refrigerator for the subsequent analysis of the expression of mRNA and protein in the small intestine; another sample was fixed in Bouin's solution for the following histological and immunohistochemical examinations.

5.3. The Concentrations of ZEA, β-ZOL and α-ZOL during Serum Detection

The serum concentrations of ZEA, α-ZOL and β-ZOL were analyzed by the Institute of Quality Standards and Detection Technology of Chinese Academy of Agricultural Sciences according to Liu et al. [28]. A LC-MS/MS analysis was performed using an Agilent 1200 liquid chromatograph (Agilent Technologies, Palo Alto, CA, USA) coupled to a 3200 QTrap® mass spectrometry system (Applied Biosystems, Foster City, CA, USA) equipped with a Turbo electrospray ionization (ESI) interface.

5.4. Small Intestine Histology Examination

The fixed small intestinal tissues were treated with a gradient series of ethanol and xylene solutions and then embedded conventionally in the paraffin wax. The embedded tissues were sliced to 5-μm-thick sections and stained with the dye of the hematoxylin and eosin (H&E staining method) to observe the small intestinal tissue structure under light microscopy. Morphometric analyses were performed using microscope analyses software (Olympus BX41, Tokyo, Japan), and the parameters, including the crypt depth, villus length and VL/CD (villus length-to-crypt depth), were measured under 40× magnification.

5.5. Immunohistochemistry (IHC)

The paraffin sections (5 μm) were dewaxed and rehydrated regularly, and the antigen retrieved was used the microwaving method for about 20 min in 0.01-mol/L, pH 6.0 sodium citrate buffer. Subsequently, the sections were sealed with 3% H_2O_2 for about 30 min in order to block the endogenous peroxidase, then incubated in 10% normal calf or goat serum (ZSGB-BIO, Beijing, China) for about 30 min to block the nonspecific binding. Hereafter, we incubated the sections with mouse anti-PCNA monoclonal antibody (1:200, ZSGB-BIO) or rabbit anti-ghrelin polyclonal antibody (1:100, BIOSS, Beijing, China) at 4 °C overnight. On the following day, the sections were washed using 0.01-mol/L, pH 7.2 PBS and were subsequently covered using the corresponding secondary antibody for 1–1.5 h at 37 °C, according to the immunohistochemical enhanced kit instructions of the labeled polymer system kit (Polink-2 plus polymer HRP detection system for mouse or rabbit specific primary antibody, PV-9002/PV-9001, ZSGB-BIO); subsequently, the sections were washed with 0.01-mol/L, pH 7.2 PBS, followed by immersion in a DAB horseradish peroxidase color development kit (DAB kit, Tiangen for about 1–3 min to detect the result of immunohistological staining. At last, the slides were counterstained with hematoxylin dye and observed the distribution of immuno-positive cells with the yellow/brownish yellow color of the immunoreactive substances under a light microscope. At the same time, the negative control tissues sections were conducted with the same program, except that the specific primary antibody was substituted with PBS and 10% goat serum.

5.6. Measurement of the Integrated Optical Density

Ghrelin and PCNA labeling were examined by a microscope (Olympus BX41 (DP25), Japan). From each specimen, views of the fields in 10 stained sections were randomly chosen (n = 10 sections, per treatment) and photographed using the Olympus microscope camera system (Olympus BX41 (DP25), Japan). The next step was using Image Pro-Plus

6.0 analysis software (Media Cybernetics, MD, Maryland, USA) to analyze and evaluate the amount of immuno-positive cells stained. The SIOD (Integrated Optical Density of the single small intestinal villi) and IOD (Total Integrated Optical Density of a cross-section) were obtained, which were applied to compare the ghrelin and PCNA staining intensities in the duodenum, jejunum and ileum between the control and ZEA treatment.

5.7. Quantitative Real-Time PCR

Total RNAs were extracted from the small intestine samples with RNAiso Plus (Lot: AA4602-1, TaKaRa, Beijing, China), according to the kit manufacturer's instructions and the literature of Dai et al. [24]. The concentration and purity of the RNA was evaluated using an Eppendorf biophotometer (Lot: RS323C, Eppendorf, Germany). The integrity of the total RNA was checked by AGE (agarose gel electrophoresis); then, we moved onto the next step, using the PrimeScript^{TM} RT Master Mix reverse transcription system kit (RR036A, TaKaRa, Beijing, China); the total RNA was reverse-transcribed onto the cDNA.

A total 20-μL volume of the qRT-PCR reaction mixture were contained with 6.8 μL of dH_2O; 0.4 μL (10 μmol/L) of forward primers and 0.4 μL (10 μmol/L) of reverse primers, 10 μL of SYBRY Premix Ex Taq II (Lot: AK7502; TaKaRa, Beijing, China), 0.4 μL of ROX reference Dye II and 2.0 μL of the cDNA (<100 ng). The protocol of qRT-PCR consisted of an initial denaturation step at 95 °C for 30 s, followed by 40 cycles for 5 s at 95 °C, for 34 s at 60 °C, for 15 s at 95 °C and for 60 s at 60 °C; the final step was for 15 s at 95 °C. The reactions were conducted in the ABI 7500 Real-Time PCR System (Applied Biosystems, Foster City, CA, USA). The ghrelin and PCNA relative mRNA quantification levels were calculated and expressed as being the $2^{-\Delta\Delta CT}$ [65] method. The qRT-PCR experiments were carried out in triplicate. In Table 4, the primer sequences of PCNA, ghrelin and GAPDH glyceraldehyde-3-phosphate dehydrogenase and the production lengths were presented.

5.8. Western Blot Analysis

The total protein of the small intestine was extracted using the Protein Extraction Kit (Beyotime, Shanghai, China) according to the kit manufacturer's instructions, and the concentration was determined using the BCA protein quantitative kit (Tiangen). An equivalent amount of protein (40 μg) was loaded onto SDS-PAGE and blotted onto a PVDF (polyvinylidene difluoride) membrane. Then, the PVDF membranes were incubated with rabbit anti-ghrelin polyclonal antibody (1:250, BIOSS), mouse anti-PCNA monoclonal antibody (1:300, ZSGB-BIO) and mouse monoclonal anti-β-actin antibody (1:3000, Beyotime) at 4 °C for 12 h. After washing with pH 7.2, 0.01-mol/L TBST, the PVDF membranes were soaked with an anti-rabbit/mouse horseradish peroxidase-labeled antibody (1:3000) for 2.5 h at room temperature. The next step was incubating the PVDF membranes with a BeyoECL Plus kit (Beyotime), followed by detection with FusionCapt Advance FX7 (Beijing Oriental Science and Technology Development Co. Ltd., Beijing, China), and were quantified using Image Pro-Plus 6.0 software (Media Cybernetics, Sliver Springs, MD, USA).

5.9. Statistical Analysis

All data were expressed as the mean ± SD. In order to confirm the differences between the treatments, the one-way ANOVA and independent-sample *t*-test of the SPSS method for Windows (version 14.0) of the analyzed data were used, with $p < 0.05$ being considered a significant difference.

Author Contributions: Conceptualization, L.H. and X.Y.; Data curation, Q.Z. and S.J.; Formal analysis, Q.Z., L.H. and S.J.; Funding acquisition, W.Y., L.H. and S.J.; Methodology, W.Y., L.H. and S.J.; Writing—original draft, Q.Z. and B.L. and Writing—review and editing, N.J., Y.L. and X.Y. All authors have read and agreed to the published version of the manuscript.

Funding: This research was financed in part by the Natural Science Foundation of Shandong Province (grant no. ZR2019MC038), Major Innovative Projects in Shandong Province of the research and application of environmentally friendly feed and critical technologies for pigs and poultry without antibiotics (grant no. 2019JZZY020609) and Founds of the Shandong Agriculture Research System in Shandong Province (grant no. SDAIT-08-04).

Institutional Review Board Statement: The authors confirm that the ethical policies of the journal, as noted on the journal's author guidelines page, have been adhered to. All animal care and experimental procedures were in accordance with the guidelines for the care and use of laboratory animals prescribed by the Animal Nutrition Research Institute of Shandong Agricultural University and the Ministry of Agriculture of China.

Informed Consent Statement: Not applicable.

Data Availability Statement: Not applicable.

Conflicts of Interest: The authors declare that there are no competing financial interests in the work described.

References

1. Obremski, K.; Zielonka, L.; Gajecka, M.; Jakimiuk, E.; Bakuła, T.; Baranowski, M.; Gajęcki, M. Histological estimation of the small intestine wall after administration of feed containing deoxynivalenol, T-2 toxin and zearalenone in the pig. *Pol. J. Vet. Sci.* **2008**, *11*, 339–345.
2. Gajecki, M.T.; Gajecka, M.; Zielonka, L. The presence of mycotoxins in feed and their influence on animal health. *Toxins* **2020**, *12*, 663. [CrossRef] [PubMed]
3. Pack, E.; Stewart, J.; Rhoads, M.; Knight, J.; De Vita, R.; Clark-Deener, S.; Schmale, D.G., III. Quantification of zearalenone and alpha-zearalenol in swine liver and reproductive tissues using GC-MS. *Toxicon X* **2020**, *8*, 100058. [CrossRef]
4. Wan, L.Y.; Turner, P.C.; El-Nezami, H. Individual and combined cytotoxic effects of Fusarium toxins (deoxynivalenol, nivalenol, zearalenone and fumonisins B1) on swine jejunal epithelial cells. *Food Chem. Toxicol.* **2013**, *57*, 276–283. [CrossRef] [PubMed]
5. Su, Y.; Sun, Y.; Ju, D.; Chang, S.; Shi, B.; Shan, A. The detoxification effect of vitamin C on zearalenone toxicity in piglets. *Ecotoxicol. Environ. Saf.* **2018**, *158*, 284–292. [CrossRef]
6. Zhang, Y.; Gao, R.; Liu, M.; Shi, B.; Shan, A.; Cheng, B. Use of modified halloysite nanotubes in the feed reduces the toxic effects of zearalenone on sow reproduction and piglet development. *Theriogenology* **2015**, *83*, 932–941. [CrossRef]
7. Cheng, Q.; Jiang, S.; Huang, L.; Ge, J.; Wang, Y.; Yang, W. Zearalenone induced oxidative stress in the jejunum in postweaning gilts through modulation of the Keap1-Nrf2 signaling pathway and relevant genes. *J. Anim. Sci.* **2019**, *97*, 1722–1733. [CrossRef] [PubMed]
8. Gajecka, M.; Zielonka, L.; Gajecki, M. Activity of zearalenone in the porcine intestinal tract. *Molecules* **2016**, *22*, 18. [CrossRef] [PubMed]
9. Marin, D.E.; Motiu, M.; Taranu, I. Food contaminant zearalenone and its metabolites affect cytokine synthesis and intestinal epithelial integrity of porcine cells. *Toxins* **2015**, *7*, 1979–1988. [CrossRef]
10. Rajendran, P.; Ammar, R.B.; Al-Saeedi, F.J.; Mohamed, M.E.; ElNaggar, M.A.; Al-Ramadan, S.Y.; Bekhet, G.M.; Soliman, A.M. Kaempferol inhibits zearalenone-induced oxidative stress and apoptosis via the PI3K/Akt-Mediated Nrf2 Signaling pathway: In vitro and in vivo Studies. *Int. J. Mol. Sci.* **2020**, *22*, 217. [CrossRef]
11. Taranu, I.; Braicu, C.; Marin, D.E.; Pistol, G.C.; Motiu, M.; Balacescu, L.; Beridan Neagoe, I.; Burlacu, R. Exposure to zearalenone mycotoxin alters in vitro porcine intestinal epithelial cells by differential gene expression. *Toxicol. Lett.* **2015**, *232*, 310–325. [CrossRef]
12. Lewczuk, B.; Przybylska-Gornowicz, B.; Gajecka, M.; Targonska, K.; Ziolkowska, N.; Prusik, M.; Gajecki, M. Histological structure of duodenum in gilts receiving low doses of zearalenone and deoxynivalenol in feed. *Exp. Toxicol. Pathol.* **2016**, *68*, 157–166. [CrossRef] [PubMed]
13. Oswald, I.P. Role of intestinal epithelial cells in the innate immune defence of the pig intestine. *Vet. Res.* **2006**, *37*, 359–368. [CrossRef] [PubMed]
14. Fan, W.; Lv, Y.; Ren, S.; Shao, M.; Shen, T.; Huang, K.; Zhou, J.; Yan, L.; Song, S. Zearalenone (ZEA)-induced intestinal inflammation is mediated by the NLRP3 inflammasome. *Chemosphere* **2018**, *190*, 272–279. [CrossRef]
15. Fan, W.; Shen, T.; Ding, Q.; Lv, Y.; Li, L.; Huang, K.; Yan, L.; Song, S. Zearalenone induces ROS-mediated mitochondrial damage in porcine IPEC-J2 cells. *J. Biochem. Mol. Toxicol.* **2017**, *31*, e21944–e21953. [CrossRef] [PubMed]
16. Perchard, R.; Clayton, P.E. Ghrelin and growth. *Endocr. Dev.* **2017**, *32*, 74–86. [CrossRef] [PubMed]
17. Mani, B.K.; Zigman, J.M. Ghrelin as a survival hormone. *Trends Endocrinol. Metab.* **2017**, *28*, 843–854. [CrossRef]
18. Lv, Y.; Liang, T.; Wang, G.; Li, Z. Ghrelin, a gastrointestinal hormone, regulates energy balance and lipid metabolism. *Biosci. Rep.* **2018**, *38*, BSR20181061. [CrossRef]
19. Yanagi, S.; Sato, T.; Kangawa, K.; Nakazato, M. The homeostatic force of ghrelin. *Cell Metab.* **2018**, *27*, 786–804. [CrossRef]
20. Eissa, N.; Ghia, J.E. Immunomodulatory effect of ghrelin in the intestinal mucosa. *Neurogastroenterol. Motil.* **2015**, *27*, 1519–1527. [CrossRef]

21. Kitazawa, T.; Kaiya, H. Regulation of gastrointestinal motility by motilin and ghrelin in vertebrates. *Front. Endocrinol.* **2019**, *10*, 278–294. [CrossRef] [PubMed]

22. Dong, X.Y.; Xu, J.; Tang, S.Q.; Li, H.Y.; Jiang, Q.Y.; Zou, X.T. Ghrelin and its biological effects on pigs. *Peptides* **2009**, *30*, 1203–1211. [CrossRef] [PubMed]

23. Willemen, S.A.; De Vos, M.; Huygelen, V.; Fransen, E.; Tambuyzer, B.R.; Casteleyn, C.; Van Cruchten, S.; Van Ginneken, C. Ghrelin in the gastrointestinal tract and blood circulation of perinatal low and normal weight piglets. *Animal* **2013**, *7*, 1978–1984. [CrossRef]

24. Dai, M.; Jiang, S.; Yuan, X.; Yang, W.; Yang, Z.; Huang, L. Effects of zearalenone-diet on expression of ghrelin and PCNA genes in ovaries of post-weaning piglets. *Anim. Reprod. Sci.* **2016**, *168*, 126–137. [CrossRef] [PubMed]

25. Kuiper-Goodman, T.; Scott, P.; Watanabe, H. Risk Assessment of the mycotoxin zearalenone. *Regul. Toxicol. Pharmacol.* **1987**, *7*, 253–306. [CrossRef]

26. Yang, J.Y.; Wang, G.X.; Liu, J.L.; Fan, J.J.; Cui, S. Toxic effects of zearalenone and its derivatives alpha-zearalenol on male reproductive system in mice. *Reprod. Toxicol.* **2007**, *24*, 381–387. [CrossRef]

27. Zinedine, A.; Soriano, J.M.; Molto, J.C.; Manes, J. Review on the toxicity, occurrence, metabolism, detoxification, regulations and intake of zearalenone: An oestrogenic mycotoxin. *Food Chem. Toxicol.* **2007**, *45*, 1–18. [CrossRef]

28. Liu, X.; Xu, C.; Yang, Z.; Yang, W.; Huang, L.; Wang, S.; Liu, F.; Liu, M.; Wang, Y.; Jiang, S. Effects of dietary zearalenone exposure on the growth performance, small intestine disaccharidase, and antioxidant activities of weaned Gilts. *Animals* **2020**, *10*, 2157. [CrossRef]

29. Antfolk, M.; Jensen, K.B. A bioengineering perspective on modelling the intestinal epithelial physiology in vitro. *Nat. Commun.* **2020**, *11*, 6244–6254. [CrossRef]

30. Avritscher, E.B.; Cooksley, C.D.; Elting, L.S. Scope and epidemiology of cancer therapy-induced oral and gastrointestinal mucositis. *Semin. Oncol. Nurs.* **2004**, *20*, 3–10. [CrossRef]

31. Billeschou, A.; Hunt, J.E.; Ghimire, A.; Holst, J.J.; Kissow, H. Intestinal adaptation upon chemotherapy-induced intestinal injury in mice depends on GLP-2 receptor activation. *Biomedicines* **2021**, *9*, 46. [CrossRef] [PubMed]

32. Jia, R.; Liu, W.; Zhao, L.; Cao, L.; Shen, Z. Low doses of individual and combined deoxynivalenol and zearalenone in naturally moldy diets impair intestinal functions via inducing inflammation and disrupting epithelial barrier in the intestine of piglets. *Toxicol. Lett.* **2020**, *333*, 159–169. [CrossRef]

33. Liew, W.P.; Mohd-Redzwan, S. Mycotoxin: Its impact on gut health and microbiota. *Front. Cell Infect. Microbiol.* **2018**, *8*, 60–76. [CrossRef]

34. Wang, P.; Huang, L.; Yang, W.; Liu, Q.; Li, F.; Wang, C. Deoxynivalenol induces inflammation in the small intestine of weaned rabbits by activating mitogen-activated protein kinase signaling. *Front. Vet. Sci.* **2021**, *8*, 632599–632608. [CrossRef]

35. Przybylska-Gornowicz, B.; Lewczuk, B.; Prusik, M.; Hanuszewska, M.; Petrusewicz-Kosinska, M.; Gajecka, M.; Zielonka, L.; Gajecki, M. The effects of deoxynivalenol and zearalenone on the pig large intestine. a light and electron microscopy Study. *Toxins* **2018**, *10*, 148. [CrossRef]

36. Altshuler, A.E.; Lamadrid, I.; Li, D.; Ma, S.R.; Kurre, L.; Schmid-Schonbein, G.W.; Penn, A.H. Transmural intestinal wall permeability in severe ischemia after enteral protease inhibition. *PLoS ONE* **2014**, *9*, e96655–e96668. [CrossRef] [PubMed]

37. Varga, J.; Tóth, Š.; Tomečková, V.; Gregová, K.; Veselá, J. The relationship between morphology and disaccharidase activity in ischemia- reperfusion injured intestine. *Acta Biochim. Pol.* **2012**, *59*, 631–638. [CrossRef] [PubMed]

38. Przybylska-Gornowicz, B.; Tarasiuk, M.; Lewczuk, B.; Prusik, M.; Ziolkowska, N.; Zielonka, L.; Gajecki, M.; Gajecka, M. The effects of low doses of two Fusarium toxins, zearalenone and deoxynivalenol, on the pig jejunum. A light and electron microscopic study. *Toxins* **2015**, *7*, 4684–4705. [CrossRef] [PubMed]

39. Biehl, M.L.; Prelusky, D.B.; Koritz, G.D.; Hartin, K.E.; Buck, W.B.; Trenholm, H.L. Biliary excretion and enterohepatic cycling of zearalenone in immature pigs. *Regul. Toxicol. Pharmacol.* **1993**, *121*, 152–159. [CrossRef]

40. Peeters, T.L. Ghrelin and the gut. *Endocr. Dev.* **2013**, *25*, 41–48. [CrossRef]

41. Zhang, S.; Okuhara, Y.; Iijima, M.; Takemi, S.; Sakata, I.; Kaiya, H.; Teraoka, H.; Kitazawa, T. Identification of pheasant ghrelin and motilin and their actions on contractility of the isolated gastrointestinal tract. *Gen. Comp. Endocrinol.* **2020**, *285*, 113294. [CrossRef]

42. Zhang, S.; Teraoka, H.; Kaiya, H.; Kitazawa, T. Motilin- and ghrelin-induced contractions in isolated gastrointestinal strips from three species of frogs. *Gen. Comp. Endocrinol.* **2021**, *300*, 113649–113682. [CrossRef]

43. Date, Y.; Kojima, M.; Hosoda, H.; Sawaguchi, A.; Mondal, M.S.; Suganuma, T.; Matsukura, S.; Kangawa, K.; Nakazato, M. Ghrelin, a novel growth hormone-releasing acylated peptide, is synthesized in a distinct endocrine cell type in the gastrointestinal tracts of rats and humans. *Endocrinology* **2000**, *141*, 4255–4261. [CrossRef]

44. Joost, O.; Scott, F.R.; Grill, H.J.; Kaplan, J.M.; Cummings, D.E. Role of the duodenum and macronutrient type in ghrelin regulation. *Endocrinology* **2005**, *146*, 845–850. [CrossRef]

45. Onishi, S.; Kaji, T.; Yamada, W.; Nakame, K.; Machigashira, S.; Kawano, M.; Yano, K.; Harumatsu, T.; Yamada, K.; Masuya, R.; et al. Ghrelin stimulates intestinal adaptation following massive small bowel resection in parenterally fed rats. *Peptides* **2018**, *106*, 59–67. [CrossRef]

46. El-Salhy, M. Ghrelin in gastrointestinal diseases and disorders: A possible role in the pathophysiology and clinical implications (review). *Int. J. Mol. Med.* **2009**, *24*, 727–732. [CrossRef]

47. Tumer, C.; Oflazoglu, H.D.; Obay, B.D.; Kelle, M.; Tasdemir, E. Effect of ghrelin on gastric myoelectric activity and gastric emptying in rats. *Regul. Pept.* **2008**, *146*, 26–32. [CrossRef]
48. Alamri, B.N.; Shin, K.; Chappe, V.; Anini, Y. The role of ghrelin in the regulation of glucose homeostasis. *Horm. Mol. Biol. Clin. Investig.* **2016**, *26*, 3–11. [CrossRef]
49. King, S.J.; Rodrigues, T.; Watts, A.; Murray, E.; Wilson, A.; Abizaid, A. Investigation of a role for ghrelin signaling in binge-like feeding in mice under limited access to high-fat diet. *Neuroscience* **2016**, *319*, 233–245. [CrossRef]
50. Cheng, Q.; Jiang, S.; Huang, L.; Wang, Y.; Yang, W.; Yang, Z.; Ge, J. Effects of zearalenone-induced oxidative stress and Keap1-Nrf2 signaling pathway-related gene expression in the ileum and mesenteric lymph nodes of post-weaning gilts. *Toxicology* **2020**, *429*, 152337–152378. [CrossRef]
51. Cheng, Y.; Wei, Y.; Yang, W.; Cai, Y.; Chen, B.; Yang, G.; Shang, H.; Zhao, W. Ghrelin attenuates intestinal barrier dysfunction following intracerebral hemorrhage in mice. *Int. J. Mol. Sci.* **2016**, *17*, 2032. [CrossRef]
52. Yamada, W.; Kaji, T.; Onishi, S.; Nakame, K.; Yamada, K.; Kawano, T.; Mukai, M.; Souda, M.; Yoshioka, T.; Tanimoto, A.; et al. Ghrelin improves intestinal mucosal atrophy during parenteral nutrition: An experimental study. *J. Pediatr. Surg.* **2016**, *51*, 2039–2043. [CrossRef]
53. Hatoya, S.; Torii, R.; Kumagai, D.; Sugiura, K.; Kawate, N.; Tamada, H.; Sawada, T.; Inaba, T. Expression of estrogen receptor α and β genes in the mediobasal hypothalamus, pituitary and ovary during the canine estrous cycle. *Neurosci. Lett.* **2003**, *347*, 131–135. [CrossRef]
54. Petruzzelli, M.; Piccinin, E.; Pinto, C.; Peres, C.; Bellafante, E.; Moschetta, A. Biliary phospholipids sustain enterocyte proliferation and intestinal tumor progression via nuclear receptor Lrh1 in mice. *Sci. Rep.* **2016**, *6*, 39278–39287. [CrossRef]
55. Phoophitphong, D.; Wangnaitham, S.; Srisuwatanasagul, S.; Tummaruk, P. The use of proliferating cell nuclear antigen (PCNA) immuno-staining technique to determine number and type of follicles in the gilt ovary. *Livest. Sci.* **2012**, *150*, 425–431. [CrossRef]
56. Liu, M.; Gao, R.; Meng, Q.; Zhang, Y.; Bi, C.; Shan, A. Toxic effects of maternal zearalenone exposure on intestinal oxidative stress, barrier function, immunological and morphological changes in rats. *PLoS ONE* **2014**, *9*, e106412–e106425. [CrossRef]
57. Fan, L.; Bi, T.; Wang, L.; Xiao, W. DNA-damage tolerance through PCNA ubiquitination and sumoylation. *Biochem. J.* **2020**, *477*, 2655–2677. [CrossRef]
58. Ripley, B.M.; Gildenberg, M.S.; Washington, M.T. Control of DNA damage bypass by ubiquitylation of PCNA. *Genes* **2020**, *11*, 138. [CrossRef]
59. Verdile, N.; Mirmahmoudi, R.; Brevini, T.A.L.; Gandolfi, F. Evolution of pig intestinal stem cells from birth to weaning. *Animal* **2019**, *13*, 2830–2839. [CrossRef]
60. Yang, L.J.; Huang, L.B.; Li, S.M.; Liu, F.X.; Jiang, S.Z.; Yang, Z.B. Effects of zearalenone on ovary index, distribution and expression of progesterone receptors in ovaries of weaned gilts. *Chin. J. Anim. Nutr.* **2017**, *29*, 4510–4517. [CrossRef]
61. Yang, L.J.; Zhou, M.; Huang, L.B.; Yang, W.R.; Yang, Z.B.; Jiang, S.Z. Zearalenone promotes follicle growth through modulation of wnt-1/β-catenin signaling pathway and expression of estrogen receptor genes in ovaries of post-weaning piglets. *J. Agric. Food Chem.* **2018**, *66*, 7899–7906. [CrossRef]
62. Song, T.T.; Liu, X.F.; Yuan, X.J.; Yang, W.R.; Liu, F.X.; Hou, Y.M.; Huang, L.B.; Jiang, S.Z. Dose-effect of zearalenone on the localization and expression of growth hormone, growth hormone receptor, and heat shock protein 70 in the ovaries of post-weaning gilts. *Front. Vet. Sci.* **2021**, *8*, 629006–629017. [CrossRef]
63. Zhou, M.; Yang, L.J.; Chen, Y.H.; Sun, T.; Wang, N.; Chen, X.; Yang, Z.B.; Ge, J.S.; Jiang, S.Z. Comparative study of stress response, growth, and development of uteri in post-weaning gilts challenged with zearalenone and estradiol benzoate. *J. Anim. Physiol. Anim. Nutr.* **2019**, *103*, 1885–1894. [CrossRef]
64. Zhou, M.; Yang, L.; Shao, M.; Wang, Y.; Yang, W.; Huang, L.; Zhou, X.; Jiang, S.; Yang, Z. Effects of zearalenone exposure on the TGF-beta1/Smad3 signaling pathway and the expression of proliferation or apoptosis related genes of post-weaning gilts. *Toxins* **2018**, *10*, 49. [CrossRef]
65. Livak, K.J.; Schmittgen, T.D. Analysis of relative gene expression data using real-time quantitative PCR and the 2 (-Delta Delta C(T)) method. *Methods* **2001**, *25*, 402–408. [CrossRef]

Article

Response of Fecal Bacterial Flora to the Exposure of Fumonisin B1 in BALB/c Mice

Fan Zhang, Zhiwei Chen, Lin Jiang, Zihan Chen and Hua Sun *

State Key Laboratory of Bioactive Substance and Function of Natural Medicines, Institute of Materia Medica, Chinese Academy of Medical Sciences & Peking Union Medical College, 1# Xiannongtan Street, Xicheng District, Beijing 100050, China; AFUN008034@163.com (F.Z.); chenzw@imm.ac.cn (Z.C.); gillianjiang@imm.ac.cn (L.J.); chenzihan@imm.ac.cn (Z.C.)
* Correspondence: sunhua@imm.ac.cn

Abstract: Fumonisins are a kind of mycotoxin that has harmful influence on the health of humans and animals. Although some research studies associated with fumonisins have been reported, the regulatory limits of fumonisins are imperfect, and the effects of fumonisins on fecal bacterial flora of mice have not been suggested. In this study, in order to investigate the effects of fumonisin B1 (FB1) on fecal bacterial flora, BALB/c mice were randomly divided into seven groups, which were fed intragastrically with 0 mg/kg, 0.018 mg/kg, 0.054 mg/kg, 0.162 mg/kg, 0.486 mg/kg, 1.458 mg/kg and 4.374 mg/kg of FB1 solutions, once a day for 8 weeks. Subsequently, feces were collected for analysis of microflora. The V3-V4 16S rRNA of fecal bacterial flora was sequenced using the Illumina MiSeq platform. The results revealed that fecal bacterial flora of mice treated with FB1 presented high diversity. Additionally, the composition of fecal bacterial flora of FB1 exposure groups showed marked differences from that of the control group, especially for the genus types including Alloprevotella, Prevotellaceae_NK3B31_group, Rikenellaceae_RC9_gut_group, Parabacteroides and phylum types including Cyanobacteria. In conclusion, our data indicate that FB1 alters the diversity and composition of fecal microbiota in mice. Moreover, the minimum dose of FB1 exposure also causes changes in fecal microbiota to some extent. This study is the first to focus on the dose-related effect of FB1 exposure on fecal microbiota in rodent animals and gives references to the regulatory doses of fumonisins for better protection of human and animal health.

Keywords: fumonisin B1; BALB/c mice; fecal bacterial flora; 16S rRNA sequencing

Key Contribution: This study is the first to focus on the dose-related effect of FB1 exposure on fecal microbiota in rodent animals and gives references to the regulatory doses of fumonisins for better protection of human and animal health.

Citation: Zhang, F.; Chen, Z.; Jiang, L.; Chen, Z.; Sun, H. Response of Fecal Bacterial Flora to the Exposure of Fumonisin B1 in BALB/c Mice. *Toxins* **2021**, *13*, 612. https://doi.org/10.3390/toxins13090612

Received: 5 July 2021
Accepted: 24 August 2021
Published: 31 August 2021

1. Introduction

Evidence has suggested that the intestinal bacterial flora plays an important role in maintaining the health of the body [1]. This community consists of trillions of bacteria, which are beneficial to nutrition collection, immunity intrusion, barrier function, and energy regulation [2–4]. However, stimulation or interference including physical destruction, xenobiotics, host factors and antimicrobial drugs, may cause changes in the numbers or species of intestinal bacterial flora [5,6], which result in diseases by affecting the proinflammatory activity [7], metabolism [8] and modulation of gut hormones [9]. Mycotoxins are secondary metabolites produced by fungi, which are capable of causing noxious effects on humans and animals [10,11]. Increased investigations on mycotoxins, such as aflatoxin B1 [12,13], deoxynivalenol [14,15], zearalenone [16], trichothecene [17], etc., have proved that exposure of mycotoxins could influence the diversity and composition of gut microbiome. Meanwhile, the intestinal bacterial flora has the ability to bind, degrade or transform mycotoxins [18,19]. Mycotoxins and intestinal bacterial flora may interact with each other.

Fumonisins are a group of mycotoxins produced by *Fusarium moniliforme* and other *Fusarium* species, which commonly contaminate corn, sorghum, related grains and even traditional Chinese medicines (TCM) throughout the world [20]. There are 28 fumonisin analogs that have been characterized since 1988 and they are separated into four main groups, identified as the fumonisin A, B, C and P series [21]. Among those many kinds of fumonisins known, fumonisin B1 (FB1) is the most abundantly produced member and the most toxic [22]. It has been proved that FB1 is neurotoxic [23], nephrotoxic, hepatotoxic [22], hepatocarcinogenic [24] and immunotoxic [25]. As a potential hazardous contaminant, FB1 has been shown to cause the production of equine leukoencephalomalacia (ELEM) and porcine pulmonary edema (PPE) [20,26]. It is also regarded as a high incidence of human esophageal cancer [26]. Contamination of food [27–30], feed [31–33] and TCM [34] with the fumonisins has been a more and more serious concern in our society. In recent years, the toxicity and the toxic mechanism of fumonisins has become a research hotspot following the aflatoxins. However, unlike aflatoxins, the toxicity of FB1 is still underestimated and the regulations associated with the guidance levels of fumonisins on food, feed or TCM are not currently exhaustive.

Previous investigations about FB1 mostly focused on the body organs' toxicity and its carcinogenicity [35]. Some researchers are committed to finding advanced methods to detect the amount of FB1 [36,37], and some are attempting to transform and remove FB1 to decrease its toxicity [38]. Few studies pay attention to the effects of FB1 on gut bacterial flora. The only two articles found were about the effect of a single dose of FB1 on gut microbiota in weaned pigs [39] and broiler chickens [40], which suggested that FB1 could disturb the normal activities of bacterial flora. The effects of different levels of FB1 on gut bacterial flora in rodents have not been investigated so far. In this study, six different concentrations of FB1 solutions, which were set according to the current regulations throughout the world, were continuously administered to female BALB/c mice, which are more sensitive to the toxicity of FB1 [41–43]. The results provide further evidence about the influence of FB1 on the fecal bacterial flora, especially low levels of FB1 for long-term exposure. This article is expected to attract more attention to the toxicity of FB1, and also hope to give references to the guidance levels of fumonisins and put forward methods to reduce the toxicity through modulating the gut bacterial flora.

2. Results

2.1. Diversity of Fecal Bacterial Communities of Mice

The rarefaction cure revealed that the measured sequence number was reasonable for further analysis. All rarefaction cures of the seven groups plateaued, which indicated that the majority of sequences were involved in the analysis process for each group (Figure 1a). Alpha diversity was applied in analyzing the complexity of species diversity for each group. Based on alpha diversity analysis, the Shannon index was identified. After 8 weeks of exposure to FB1, the Shannon diversity index tended to be higher in feces from FB1-treated groups, especially in FB1-3, FB1-4 and FB1-5 groups, compared to those in the control group. Unusually, FB1-6 was slightly decreased relative to other groups but still increased compared with control group. These results suggested that FB1 exposure was accompanied by higher diversity in feces microbiota (Figure 1b). Beta diversity was used to evaluate differences in the species complexity in samples and was showed by a weighted unifrac index. Only the fecal bacterial flora of FB1-3 was significantly increased compared with the control group. Based on the abundance of every OTU (Figure 1c), the data distribution of groups was exhibited by PCA (Principal Component Analysis), which indicated that FB1-3, FB1-4, FB1-5 and FB1-6 groups were separated well from the control group (Figure 2).

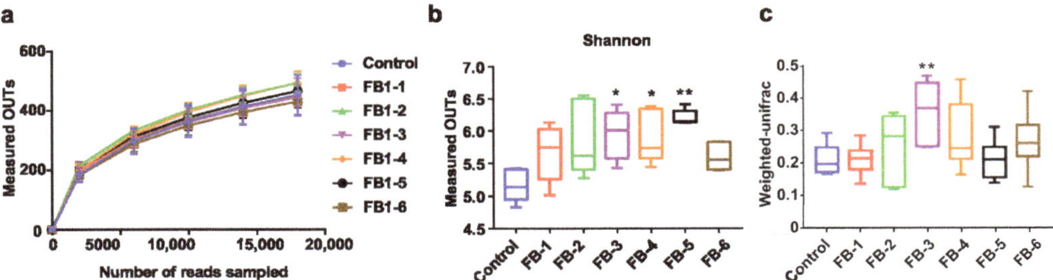

Figure 1. The diversity of the fecal bacterial community. (**a**) Rarefaction curve. (**b**) Alpha diversity revealed by the Shannon index. (**c**) Beta diversity revealed by weighted unifrac distance. * means $p < 0.05$, ** means $p < 0.01$ vs. control group.

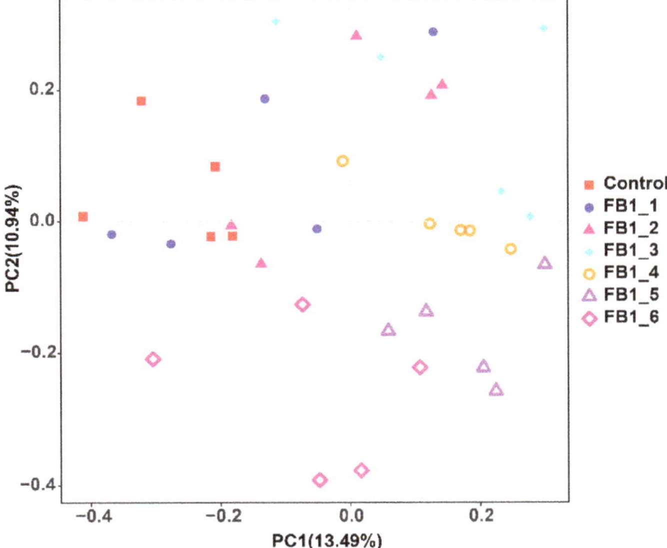

Figure 2. The PCA of the bacterial flora in groups.

2.2. Microbiota Profiles in Feces of Mice

The relative abundances of fecal bacteria in genus types were drawn as a column picture and a heatmap as shown in Figures 3–5. The 10 most abundant genera were Alloprevotella, Lactobacillus, Alistipes, Lachnospiraceae_NK4A136_group, Bacteroides, Prevotellaceae_UCG-001, Ruminococcaceae_UCG-014, Roseburia, Lachnoclostridium and Prevotellaceae_NK3B31_group, respectively. Concerning the relative abundances of fecal bacteria in phylum types, it was obtained that the Bacteroides phylum was the most abundant one in the fecal microbiota from all groups. Firmicutes was the second most abundant phylum followed by Proteobacteria and Cyanobacteria. Furthermore, Cyanobacteria tended to be higher in all FB1-treated groups than in the control group. There were also another four phylum types with relatively lower abundance (<1%) in the heatmap: Actinobacteria, Saccharibacteria, Deferribacteria and tenericutes.

2.3. Phylogenetic Tree

The phylogenetic tree was obtained based on the hierarchical clustering analysis according to the distance matrix in the beta diversity analysis (Figure 6). The fecal bacteria of the control and the low-dose groups including FB1-1 and FB1-2 were clustered together, which indicates that these three groups had a relatively close genetic relationship. In addition, FB1-3 and FB1-4, and FB1-5 and FB1-6 were clustered together, respectively, showing a close genetic relationship.

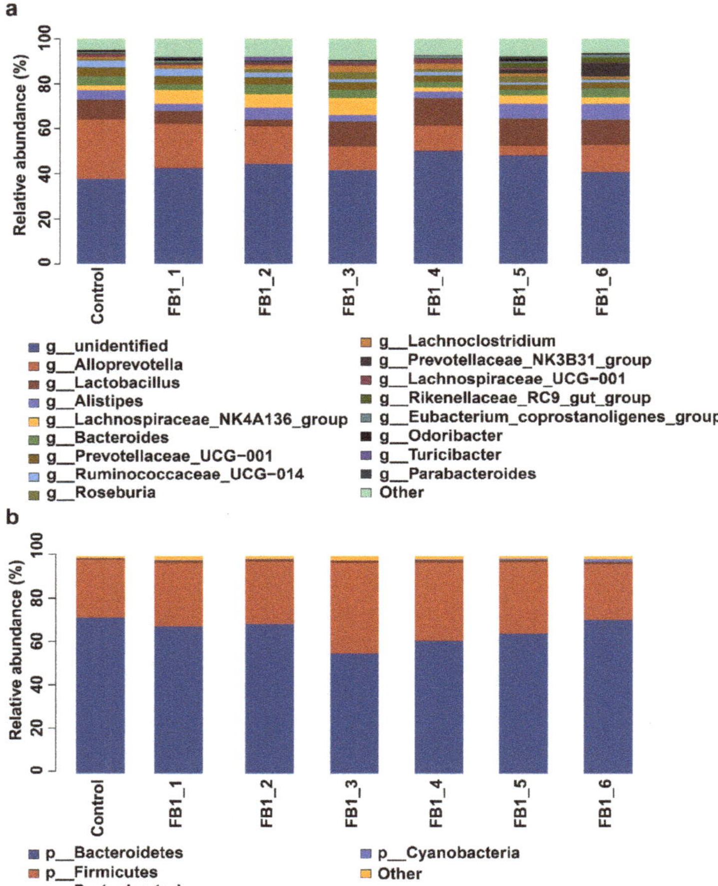

Figure 3. Column pictures of genus and phylum types and relative abundance of fecal bacterial flora. Others: Fecal bacterial flora with relative abundance < 1% were included as others. (**a**) Column pictures of genera; (**b**) column pictures of phyla.

2.4. Shifts in Relative Abundance of Fecal Bacterial Flora

2.4.1. Genera Performances

As shown in Figure 7, compared with the control group, Alloprevotella decreased significantly in FB1-3, FB1-4, FB1-5 and FB1-6 groups. Prevotallaceae_UCG-001 and Ruminococcaceae_UCG-014 tended to decrease in FB1-exposed groups compared with the control group. In contrast, Prevotellaceae_NK3B31_group, Rikenellaceae_RC9_gut_group and Parabacteroides showed a significant increase in the FB1-5 group and the FB1-6 group. In addition, there was an interesting phenomenon that Lactobacillus and Alistipes decreased in lower doses first and increased in higher doses, while Lachnospiraceae_NK4A136_group, Lachnoclostridium and Lachnospiraceae_UCG-001 typically increased in lower doses and decreased in higher doses.

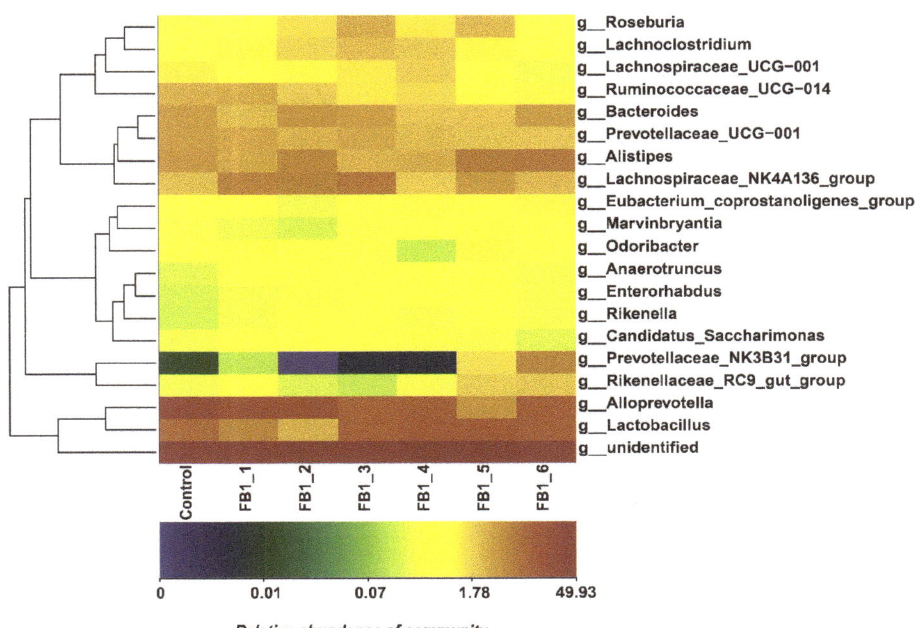

Figure 4. Heatmap of genus types and relative abundance of fecal bacterial flora. Others: Fecal bacterial flora with relative abundance < 1% were included as others.

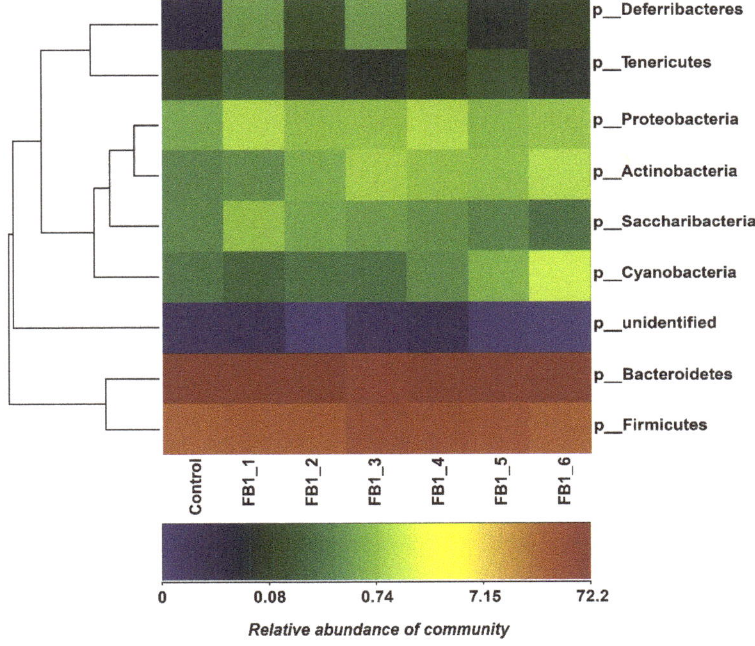

Figure 5. Heatmap of phylum types and relative abundance of fecal bacterial flora. Others: Fecal bacterial flora with relative abundance <1% were included as others.

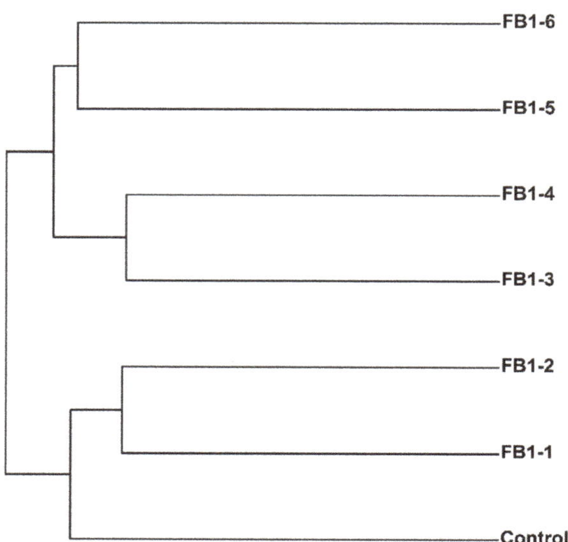

Figure 6. Phylogenetic tree obtained based on the hierarchical clustering analysis.

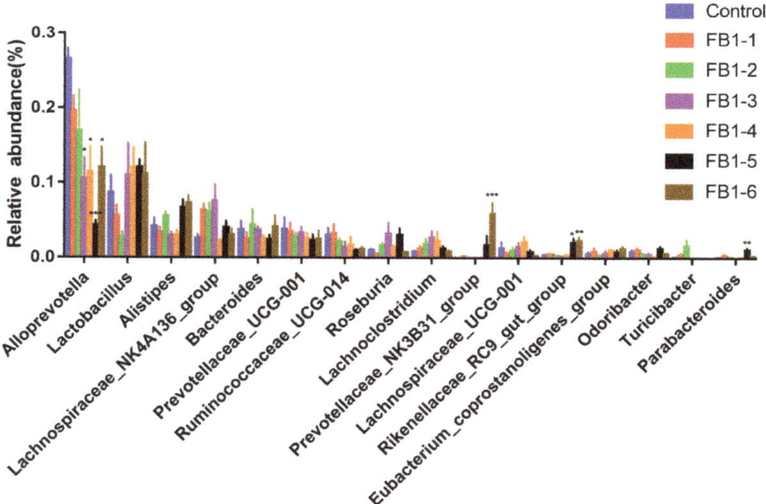

Figure 7. Difference in relative abundance of fecal bacterial flora between control and FB1 treatment groups at the genus level. * means $p < 0.05$, ** means $p < 0.01$, *** means $p < 0.001$ vs. control group.

2.4.2. Phyla Performances

As for phyla types, Bacteroidetes and Firmicutes presented opposite tends (Figure 8). We determined that Bacteroidetes decreased in FB1-1, FB1-2 and FB1-3 groups compared to the control group and increased in FB1-4, FB1-5 and FB1-6 groups, while Firmicutes showed increases in FB1-1, FB1-2 and FB1-3 groups and decreases in FB1-4, FB1-5 and FB1-6 groups. For Proteobacteria, no evident change occurred in all FB1 exposure groups. The relative abundance of Cyanobacteria showed a significant increase in the FB1-6 group compared to the control group.

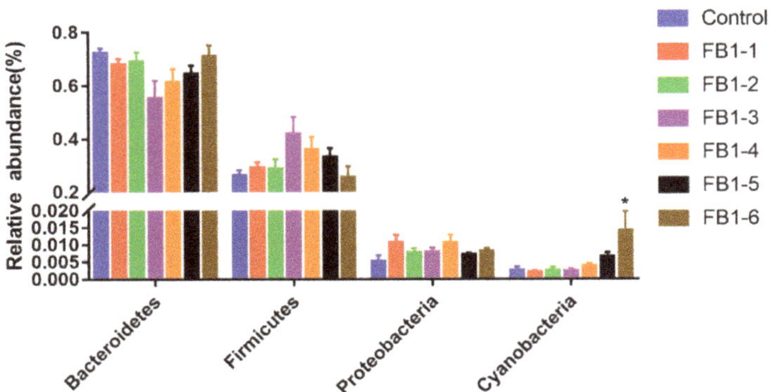

Figure 8. Difference in relative abundance of fecal bacterial flora between control and treatment groups at the phylum level. * means $p < 0.05$ vs. control group.

2.5. LEfSe Analysis

LEfSe analysis was also performed with an LDA score of three to evaluate the community structure variance of bacterial flora between the seven groups. Compared with FB1 treatments, Figure 9 clearly showed that Alloprevotella remained the absolute predominant bacteria in the control group. There were six genera that had an obvious increase in FB1 treatments (FB1-3, FB1-4, FB1-5 and FB1-6 groups): Roseburia, Lachnospiraceae_NK4A136_group, Lactobacillus, Rikenellaceae_RC9_gut_group, Alistipes and Prevotellaceae_NK3B31_group. Additionally, there were no significant changes in FB1-1 and FB1-2 groups.

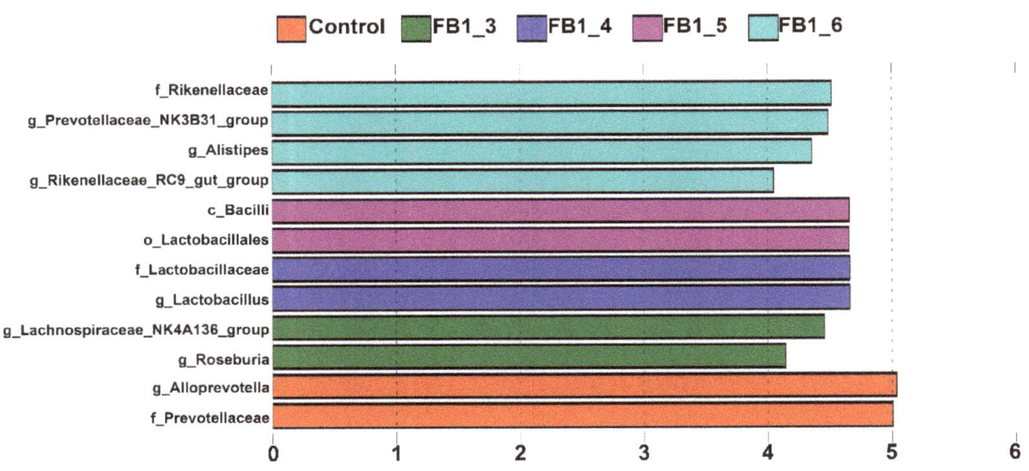

Figure 9. LEfSe comparison of fecal bacterial flora in seven groups.

2.6. Microbial Function Prediction

In order to evaluate the effects of FB1 on the fecal bacterial flora of mice, the KEGG database was used to compare and predict the function and metabolic pathways based on the microbiota flora in these seven groups. Through comparing the control with each FB1-treated group, we made some interesting observations. Both FB1-1 and FB1-2 groups inhibit the Ubiquitin system significantly. All of FB1-2, FB1-3, FB1-4 and FB1-5 groups presented obvious inhibitory effects on Glycosphingolipid biosynthesis. In addition, electron transfer carriers were obviously elevated in FB1-3, FB1-4 and FB1-5 groups. Only in the FB1-6 group did the PPAR signaling pathway, Lipopolysaccharide biosynthesis proteins and Lipopolysaccharide biosynthesis significantly increase, while Cytoskeleton proteins showed a significant decrease (Figure 10).

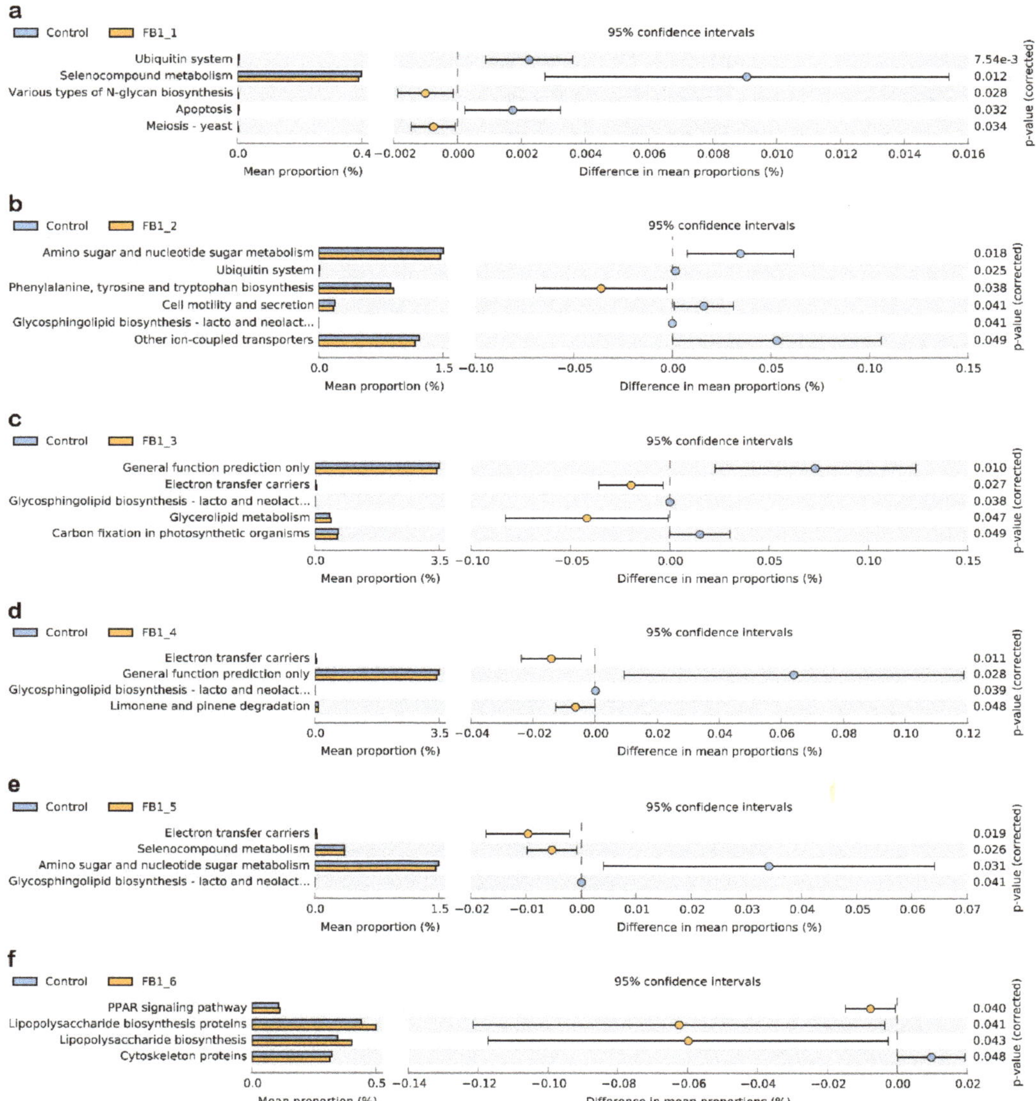

Figure 10. KEGG pathway analysis for fecal bacterial flora in seven groups. (**a**) Control group vs. FB1-1 group. (**b**) Control group vs. FB1-2 group. (**c**) Control group vs. FB1-3 group. (**d**) Control group vs. FB1-4 group. (**e**) Control group vs. FB1-5 group. (**f**) Control group vs. FB1-6 group.

3. Discussion

It has been highly agreeable that microflora in animal gut tracts plays an important role in host health. Many immune and metabolic diseases, such as chronic inflammation, obesity, diabetes, inflammatory bowel disease and atherosclerosis, are likely to be related to the imbalance of gut bacterial flora [3,5,44,45]. Paying more attention to the balance of intestinal flora is more and more urgent. Like other mycotoxins, fumonisin has detrimental effects on the health of human beings and livestock. Many papers have demonstrated that FB1 could increase the ratio of Sa/So, decrease villi length and cause obvious hepatoxicity and nephrotoxicity [20,35,46,47]. To our knowledge, this study is the first to investigate and demonstrate the dose-related effect of FB1 exposure on fecal microbiota in rodent animals. Furthermore, we also intended to observe the effect of low dosage of FB1 that corresponds to the regulatory limit on fecal bacterial flora.

Intestinal microbiota plays a crucial role in human physiology and metabolism due to its multiple functions. In this study, it was observed that FB1 treatments caused many changes in the abundance and composition of gut microflora. Treatment with FB1 significantly decreased the abundance of Alloprevotella (in FB1-3, FB1-4, FB1-5 and FB1-6 groups) and increased Prevotellaceae_NK3B31_group (in FB1-6 group), Rikenellaceae_RC9_gut_group (in FB1-5 and FB1-6 groups) and Parabacteroides (in FB1-5 group). Many research studies had reported that Alloprevotella was associated with the production of short-chain fatty acids (SCFA) and that Alloprevotella mainly produce succinate and acetate, which could protect the intestinal mucosal barrier and inhibit inflammation [48–50]. In our results, the abundance of Alloprevotella is negatively correlated with FB1 exposure. Further, it is crucial that we found that the abundance of Alloprevotella in the low dose of FB1 (FB1-1) already showed an obvious decreased tendency, which informed us to pay more attention to the PMTDI of 2 µg/kg bw for FB1, FB2 and FB3, alone or in combination. What is more, our data showed that the dose of FB1-4 below the regulatory limit of 4 mg/kg from USFDA presented a significant decrease of Alloprevotella. This suggested that the regulatory limit from USFDA may be not protective enough. Rikenellaceae has been reported to be related to the pathological progression of inflammatory bowel disease [51] and circadian rhythm disorders, especially in female mice [52]. In our study, we found that Rikenellaceae rose after FB1 treatment. All the evidence indicated that the toxicity of FB1 may be closely related to the chronic inflammation in animals. Consistent with the literature, the dominant phyla of mice were Firmicutes and Bacteroidetes [53]. We determined that exposure of FB1 did not significantly change the abundance of Firmicutes, Bacteroidetes and Proteobacteria. Our results also found that FB1 significantly increased the Cyanobacteria phylum in the FB1-6 group. Little is known about the functions of Cyanobacteria and the influence of Cyanobacteria on animal intestinal microflora [16,54]. Some research found that some species in Cyanobacteria could produce toxic metabolites known as cyanotoxins [55]. Moreover, the toxicity of cyanotoxins is strain-specific [56]. Therefore, positive identification does not predict the hazard level, that is to say, it is difficult to speculate the consequences of the increase of Cyanobacteria caused by FB1, and more studies are needed.

The phylogenetic tree reflects the similarity and difference relationship among different samples. From our phylogenetic tree results, these seven groups have been distinguished well, which indicated that FB1 may present a progressive damage effect on mice along with the increase of doses. Furthermore, the low dosage of FB1 also caused the obvious changes of fecal bacterial flora in mice. It reminded us that we should think more about the long-term accumulation of lower dose of FB1 in body. Additionally, intestinal microbiota is involved in many functional pathways through which the health of host is influenced [57]. In order to estimate how the intestinal microbiota changed after FB1 exposure, the KEGG database was used to compare the sequencing data of these groups. Based on PICRUSt analysis, the results showed that some different pathways were investigated, including the Ubiquitin system, Glycosphingolipid biosynthesis, Electron transfer carriers, PPAR signaling pathway, Lipopolysaccharide biosynthesis proteins, Lipopolysaccharide biosynthesis and Cytoskeleton proteins, etc. From the results, a low dose of FB1 (FB1-1) exposure has caused significant changes in the Ubiquitin system, Selenocompound metabolism, N-glycan biosynthesis, Apoptosis and Meiosis. Ubiquitination is a posttranslational modification process, and more and more evidence indicate the importance of ubiquitination in the immune system's response [58]. Selenium is an essential trace element and is important for various aspects of human health. Moreover, its biological effects are dependent on its incorporation into selenoproteins, which are involved in the modulation of the immune response, inflammatory processes as well as chemoprevention [59,60]. Besides, the N-glycan has influence on the transport of glycosylated proteins, and the N-glycan-dependent enzyme complex may link the processing of N-glycosylated glycosyl-transferases with glycosphingolipid metabolism [61], which presented visible alteration in FB1-2, FB1-3, FB1-4 and FB1-5-treated groups. It is reported that glycosphingolipids have specific functional roles essential for survival, proliferation and differentiation during brain development [62], which may be the likely cause of the induction of ELEM by FB1. Further related research is underway by our team. What is more, the high dose of FB1 (FB1-6) exposure caused significant changes in the PPAR signaling pathway, Lipopolysaccharide biosynthesis proteins, Lipopolysaccharide biosynthesis and Cytoskeleton proteins. The pathway of Lipopolysaccharide biosynthesis may cause the formation of numerous lipopolysaccharides (LPS), which is the main constituent of the outer membrane of most of the Gram-negative bacteria and can initiate a strong immune response and serves as an early warning signal of bacterial infection [63–65]. From these metabolic profiles analyses, we found that there may be an interesting connection or progressive change from FB1-1 to FB1-6: Ubiquitin system–Glycosphingolipid biosynthesis–Inflammation. All of the above evidence suggested that FB1 exposure could finally induce wide inflammation in mice through influencing many other metabolism pathways.

Since there are few research studies reported that are associated with the effects of FB1 on intestinal bacterial flora in animals such as weaned pigs [39] and broiler chickens [40], more investigations are needed to improve the acknowledgement of FB1.

4. Materials and Methods

4.1. FB1 Solution Preparation

The FB1 solution was prepared by dissolving FB1 powder (Pribolab, Qingdao, China) in distilled water. First of all, we obtained the 0.4374 mg/mL FB1 solution and then diluted it with distilled water to reach the concentrations of 0.1458 mg/mL, 0.0486 mg/mL, 0.0162 mg/mL, 0.0054 mg/mL and 0.0018 mg/mL.

4.2. Animal Trial

In this study, female BALB/c mice (SPF grade, HFK BIOSCIENCE Co., LTD. Beijing, China), 18–20 g body weight with no specific pathogens were used. The mice were acclimated for one week with non-restricted access to commercial feed and water. They were maintained in the environment with $20 \pm 3\,^{\circ}C$ temperature and $50.0 \pm 10.0\%$ humidity and a 12 h light/dark cycle. Animal care and all experimental procedures were approved by the Animal Ethics Committee at PUMC&CAMS and were conducted in accordance with the health criteria for the care of laboratory animals enacted by the Beijing municipal government. The mice were then randomly divided into 7 groups, which were labeled as the control, FB1-1 (0.018 mg/kg), FB1-2 (0.054 mg/kg), FB1-3 (0.162 mg/kg), FB1-4 (0.486 mg/kg), FB1-5 (1.458 mg/kg) and FB1-6 (4.374 mg/kg). The dose of FB1-1 corresponds to the PMTDI (provisional maximum tolerable daily intake) of 2 µg/kg bw for FB1, FB2 and FB3, alone or in combination by JECFA (JOINT FAO/WHO EXPERT COMMITTEE ON FOOD ADDITIVES) [66]. In addition, the dose of FB1-4 and FB1-5 crossed over the recommended levels of 4 mg/kg for total fumonisins (FB1 + FB2 + FB3) in whole or partially degermed dry milled corn products (e.g., flaking grits, corn grits, corn meal, corn flour with fat content of >2.25%, dry weight basis) by USFDA [67]. The control group was fed intragastrically with distilled water, while other groups were fed intragastrically with the corresponding FB1 solutions, respectively. The feeding dosage was 10 mL/kg per mouse each time, once a day for 8 weeks.

4.3. Sample Collection

Fresh feces were collected from the mice after 8 weeks of exposure to FB1 and then placed in liquid nitrogen at once and stored at $-80\,°C$.

4.4. 16S rRNA Gene Sequencing of Fecal Bacterial Flora

The total genomic DNA was extracted from fecal samples by using the PowerSoil DNA Isolation Kit (MoBio Laboratories, Carlsbad, CA, USA) according to the manufacturer's instructions. The quality and concentrations of DNA were measured by the NanoDrop ND-1000 Spectrophotometer (NanoDrop Technologies, Wilmington, DE, USA). The V3-V4 hypervariable region of the 16S rRNA genes were amplified and purified using the PCR primers with barcode. The sequencing library was quantified by Qubit and qPCR, and the barcoded V3-V4 PCR amplicons were sequenced using the MiSeq platform (Illumina, San Diego, CA, USA) in the double-ended sequencing mode according to the manufacturer's instructions.

The obtained paired-end reads were merged into tags, and the sequences used in the subsequent analysis were obtained by splicing raw sequence reads using FLASH (Version 1.20) software [68] and eliminating chimeric sequences using USEARCH (Version 10.0.240) software.

4.5. Taxonomy Classification and Data Analysis

Sequences with 97% similarity were assigned to the same operational taxonomic unit (OTU) using UPARSE software [69], and taxonomic annotation was conducted using the RDP classifier. The OTU abundance information was normalized using a standard of sequence number corresponding to the sample with the fewest sequences. The alpha diversity was conducted with Quantitative Insights Into Microbial Ecology (QIIME, Version v.1.8.0, http://qiime.org/, accessed on 25 August 2021) software, which is usually used to assess the diversity and abundance of microbes. The unifrac distance was obtained by QIIME (Version 1.8.0, http://qiime.org/, accessed on 25 August 2021) software, and PCA (Principal Component Analysis) was performed using R software. The LEfSe analysis was performed to obtain differences in the abundance, and the threshold of LDA score was 3.0. Besides, in this study, the KEGG database was used to annotate microbiome genes and predict the metabolic function of microbiota with the PICRUSt (Phylogenetic Investigation of Communities by Reconstruction of Unobserved States) program (http://picrust.github.io/picrust/, accessed on 25 August 2021) [70].

4.6. Statistical Analysis

Statistical significance was determined following the test using either a one-tailed t-test or one-way analysis of variance (ANOVA) with Tukey's multiple comparison to compare the means of multiple groups. Data are shown as mean $\pm$ SEM and were considered statistically significant at $p < 0.05$. GraphPad Prism 7 (GraphPad Software Inc., LaJolla, CA, USA) was used for analysis.

5. Conclusions

This study investigated the influences of exposure to different doses of FB1 on fecal bacterial flora in female BALB/c mice. The results indicate that exposure to FB1 in mice shifts the structure and composition of the fecal bacterial community. The apparent changes appear in genus types including Alloprevotella, Prevotellaceae_NK3B31_group, Rikenellaceae_RC9_gut_group, Parabacteroides and phylum types including Cyanobacteria. Moreover, there are some indicated pathways influenced by FB1. To some degree, the minimum dose of FB1 (FB1-1, corresponding to the PMTDI of 2 µg/kg bw for FB1, FB2 and FB3, alone or in combination by JECFA) in our study also presented potential influences on fecal bacterial flora. Subsequently, our team will analyze the changes in biochemical indicators and histopathology to confirm the effects of sequence concentration

FB1 in mice. Moreover, the depth of toxic mechanisms and the mechanisms of influence on fecal microbiota are the focus of our next study on FB1.

Author Contributions: H.S. and F.Z. conceived and designed the experiments; F.Z., L.J., Z.C. (Zhiwei Chen) and Z.C. (Zihan Chen) performed the experiments; H.S., Z.C. (Zhiwei Chen) and F.Z. analyzed the data and wrote the paper. All authors have read and agreed to the published version of the manuscript.

Funding: This work was funded by the CAMS Innovation Fund for Medical Sciences (grant number 2017-I2M-1-013), the National Key Research and Development Program of China [No. 2019YFC1708901], the National Science & Technology Major Project "Key New Drug Creation and Manufacturing Program" (grant number 2019ZX09201001), and the Drug Innovation Major Project (grant number 2018ZX09711001-003-011).

Institutional Review Board Statement: The study was conducted according to the guidelines of the Declaration of Helsinki, and approved by The Animal Care & walfare Committe Instltute of Materica Medica, CAMS & PUMC. (protocol code 00003407. Approval Date: 30 October 2018).

Data Availability Statement: Data available on request due to restrictions e.g., privacy or ethical.

Conflicts of Interest: The authors declare no conflict of interest.

References

1. Sommer, F.; Bäckhed, F. The gut microbiota-masters of host development and physiology. *Nat. Rev. Microbiol.* **2013**, *11*, 227–238. [CrossRef]
2. Blaut, M.; Clavel, T. Metabolic diversity of the intestinal microbiota: Implications for health and disease. *J. Nutr.* **2007**, *137*, 751s–755s. [CrossRef] [PubMed]
3. DuPont, A.W.; DuPont, H.L. The intestinal microbiota and chronic disorders of the gut. *Nat. Rev. Gastroenterol. Hepatol.* **2011**, *8*, 523–531. [CrossRef]
4. Torres-Fuentes, C.; Schellekens, H.; Dinan, T.G.; Cryan, J.F. The microbiota-gut-brain axis in obesity. *Lancet Gastroenterol. Hepatol.* **2017**, *2*, 747–756. [CrossRef]
5. Claesson, M.J.; Jeffery, I.B.; Conde, S.; Power, S.E.; O'Connor, E.M.; Cusack, S.; Harris, H.M.; Coakley, M.; Lakshminarayanan, B.; O'Sullivan, O.; et al. Gut microbiota composition correlates with diet and health in the elderly. *Nature* **2012**, *488*, 178–184. [CrossRef] [PubMed]
6. Yadav, M.; Verma, M.K.; Chauhan, N.S. A review of metabolic potential of human gut microbiome in human nutrition. *Arch. Microbiol.* **2018**, *200*, 203–217. [CrossRef]
7. Arimatsu, K.; Yamada, H.; Miyazawa, H.; Minagawa, T.; Nakajima, M.; Ryder, M.I.; Gotoh, K.; Motooka, D.; Nakamura, S.; Iida, T.; et al. Oral pathobiont induces systemic inflammation and metabolic changes associated with alteration of gut microbiota. *Sci. Rep.* **2014**, *4*, 4828. [CrossRef] [PubMed]
8. Lynch, S.V.; Pedersen, O. The Human Intestinal Microbiome in Health and Disease. *N. Engl. J. Med.* **2016**, *375*, 2369–2379. [CrossRef] [PubMed]
9. Khosravi, Y.; Seow, S.W.; Amoyo, A.A.; Chiow, K.H.; Tan, T.L.; Wong, W.Y.; Poh, Q.H.; Sentosa, I.M.; Bunte, R.M.; Pettersson, S.; et al. Helicobacter pylori infection can affect energy modulating hormones and body weight in germ free mice. *Sci. Rep.* **2015**, *5*, 8731. [CrossRef]
10. Liew, W.P.; Mohd-Redzwan, S. Mycotoxin: Its Impact on Gut Health and Microbiota. *Front. Cell. Infect. Microbiol.* **2018**, *8*, 60. [CrossRef]
11. Agriopoulou, S.; Stamatelopoulou, E.; Varzakas, T. Advances in Occurrence, Importance, and Mycotoxin Control Strategies: Prevention and Detoxification in Foods. *Foods* **2020**, *9*, 137. [CrossRef]
12. Yang, X.; Liu, L.; Chen, J.; Xiao, A. Response of Intestinal Bacterial Flora to the Long-term Feeding of Aflatoxin B1 (AFB1) in Mice. *Toxins* **2017**, *9*, 317. [CrossRef]
13. Zhou, J.; Tang, L.; Wang, J.S. Assessment of the adverse impacts of aflatoxin B(1) on gut-microbiota dependent metabolism in F344 rats. *Chemosphere* **2019**, *217*, 618–628. [CrossRef]
14. Liao, Y.; Peng, Z.; Chen, L.; Nüssler, A.K.; Liu, L.; Yang, W. Deoxynivalenol, gut microbiota and immunotoxicity: A potential approach? *Food Chem. Toxicol.* **2018**, *112*, 342–354. [CrossRef] [PubMed]
15. Wang, S.; Yang, J.; Zhang, B.; Zhang, L.; Wu, K.; Yang, A.; Li, C.; Wang, Y.; Zhang, J.; Qi, D. Potential Link between Gut Microbiota and Deoxynivalenol-Induced Feed Refusal in Weaned Piglets. *J. Agric. Food Chem.* **2019**, *67*, 4976–4986. [CrossRef] [PubMed]
16. Li, P.; Yang, S.; Zhang, X.; Huang, S.; Wang, N.; Wang, M.; Long, M.; He, J. Zearalenone Changes the Diversity and Composition of Caecum Microbiota in Weaned Rabbit. *BioMed Res. Int.* **2018**, *2018*, 3623274. [CrossRef]
17. Gratz, S.W.; Dinesh, R.; Yoshinari, T.; Holtrop, G.; Richardson, A.J.; Duncan, G.; MacDonald, S.; Lloyd, A.; Tarbin, J. Masked trichothecene and zearalenone mycotoxins withstand digestion and absorption in the upper GI tract but are efficiently hydrolyzed by human gut microbiota in vitro. *Mol. Nutr. Food Res.* **2017**, *61*, 1600680. [CrossRef]

18. Wang, J.; Tang, L.; Glenn, T.C.; Wang, J.S. Aflatoxin B1 Induced Compositional Changes in Gut Microbial Communities of Male F344 Rats. *Toxicol. Sci.* **2016**, *150*, 54–63. [CrossRef]

19. Young, J.C.; Zhou, T.; Yu, H.; Zhu, H.; Gong, J. Degradation of trichothecene mycotoxins by chicken intestinal microbes. *Food Chem. Toxicol.* **2007**, *45*, 136–143. [CrossRef] [PubMed]

20. Tsunoda, M.; Sharma, R.P.; Riley, R.T. Early fumonisin B1 toxicity in relation to disrupted sphingolipid metabolism in male BALB/c mice. *J. Biochem. Mol. Toxicol.* **1998**, *12*, 281–289. [CrossRef]

21. Rheeder, J.P.; Marasas, W.F.; Vismer, H.F. Production of fumonisin analogs by Fusarium species. *Appl. Environ. Microbiol.* **2002**, *68*, 2101–2105. [CrossRef]

22. Bhandari, N.; Raghubir, P.; Sharma. Modulation of selected cell signaling genes in mouse liver by fumonisin B1. *Chem. Biol. Interact.* **2002**, *139*, 317–331. [CrossRef]

23. Domijan, A.M. Fumonisin B(1): A neurotoxic mycotoxin. *Arh. Hig. Rada Toksikol.* **2012**, *63*, 531–544. [CrossRef] [PubMed]

24. Fukuda, H.; Shima, H.; Vesonder, R.F.; Tokuda, H.; Nishino, H.; Katoh, S.; Tamura, S.; Sugimura, T.; Nagao, M. Inhibition of protein serine/threonine phosphatases by fumonisin B1, a mycotoxin. *Biochem. Biophys. Res. Commun.* **1996**, *220*, 160–165. [CrossRef] [PubMed]

25. Dombrink-Kurtzman, M.A.; Gomez-Flores, R.; Weber, R.J. Activation of rat splenic macrophage and lymphocyte functions by fumonisin B1. *Immunopharmacology* **2000**, *49*, 401–409. [CrossRef]

26. Dutton, M.F. Fumonisins, mycotoxins of increasing importance: Their nature and their effects. *Pharmacol. Ther.* **1996**, *70*, 137–161. [CrossRef]

27. Cendoya, E.; Nichea, M.J.; Monge, M.P.; Sulyok, M.; Chiacchiera, S.M.; Ramirez, M.L. Fumonisin occurrence in wheat-based products from Argentina. *Food Addit. Contam. Part B Surveill.* **2019**, *12*, 31–37. [CrossRef]

28. Jiang, D.; Li, F.; Zheng, F.; Zhou, J.; Li, L.; Shen, F.; Chen, J.; Li, W. Occurrence and dietary exposure assessment of multiple mycotoxins in corn-based food products from Shandong, China. *Food Addit. Contam. Part B Surveill.* **2019**, *12*, 10–17. [CrossRef]

29. Ponce-García, N.; Serna-Saldivar, S.O.; Garcia-Lara, S. Fumonisins and their analogues in contaminated corn and its processed foods—A review. *Food Addit. Contam. Part A Chem. Anal. Control Expo. Risk Assess.* **2018**, *35*, 2183–2203. [CrossRef]

30. Torović, L. Fusarium toxins in corn food products: A survey of the Serbian retail market. *Food Addit. Contam. Part A Chem. Anal. Control Expo. Risk Assess.* **2018**, *35*, 1596–1609. [CrossRef]

31. Akinmusire, O.O.; El-Yuguda, A.D.; Musa, J.A.; Oyedele, O.A.; Sulyok, M.; Somorin, Y.M.; Ezekiel, C.N.; Krska, R. Mycotoxins in poultry feed and feed ingredients in Nigeria. *Mycotoxin Res.* **2019**, *35*, 149–155. [CrossRef]

32. Seo, D.G.; Phat, C.; Kim, D.H.; Lee, C. Occurrence of Fusarium mycotoxin fumonisin B1 and B2 in animal feeds in Korea. *Mycotoxin Res.* **2013**, *29*, 159–167. [CrossRef] [PubMed]

33. Shao, M.; Li, L.; Gu, Z.; Yao, M.; Xu, D.; Fan, W.; Yan, L.; Song, S. Mycotoxins in commercial dry pet food in China. *Food Addit. Contam. Part B Surveill.* **2018**, *11*, 237–245. [CrossRef]

34. Han, Z.; Ren, Y.; Liu, X.; Luan, L.; Wu, Y. A reliable isotope dilution method for simultaneous determination of fumonisins B1, B2 and B3 in traditional Chinese medicines by ultra-high-performance liquid chromatography-tandem mass spectrometry. *J. Sep. Sci.* **2010**, *33*, 2723–2733. [CrossRef]

35. Szabó, A.; Szabó-Fodor, J.; Kachlek, M.; Mézes, M.; Balogh, K.; Glávits, R.; Ali, O.; Zeebone, Y.Y.; Kovács, M. Dose and Exposure Time-Dependent Renal and Hepatic Effects of Intraperitoneally Administered Fumonisin B$_1$ in Rats. *Toxins* **2018**, *10*, 465. [CrossRef] [PubMed]

36. Ling, S.; Wang, R.; Gu, X.; Wen, C.; Chen, L.; Chen, Z.; Chen, Q.A.; Xiao, S.; Yang, Y.; Zhuang, Z.; et al. Rapid detection of fumonisin B1 using a colloidal gold immunoassay strip test in corn samples. *Toxicon* **2015**, *108*, 210–215. [CrossRef] [PubMed]

37. Li, L.; Chen, W.; Li, H.; Iqbal, J.; Zhu, Y.; Wu, T.; Du, Y. Rapid determination of fumonisin (FB(1)) by syringe SPE coupled with solid-phase fluorescence spectrometry. *Spectrochim. Acta A Mol. Biomol. Spectrosc.* **2020**, *226*, 117549. [CrossRef] [PubMed]

38. Sharma, R.P.; Bhandari, N.; He, Q.; Riley, R.T.; Voss, K.A. Decreased fumonisin hepatotoxicity in mice with a targeted deletion of tumor necrosis factor receptor 1. *Toxicology* **2001**, *159*, 69–79. [CrossRef]

39. Mateos, I.; Combes, S.; Pascal, G.; Cauquil, L.; Barilly, C.; Cossalter, A.M.; Laffitte, J.; Botti, S.; Pinton, P.; Oswald, I.P. Fumonisin-Exposure Impairs Age-Related Ecological Succession of Bacterial Species in Weaned Pig Gut Microbiota. *Toxins* **2018**, *10*, 230. [CrossRef]

40. Antonissen, G.; Croubels, S.; Pasmans, F.; Ducatelle, R.; Eeckhaut, V.; Devreese, M.; Verlinden, M.; Haesebrouck, F.; Eeckhout, M.; De Saeger, S.; et al. Fumonisins affect the intestinal microbial homeostasis in broiler chickens, predisposing to necrotic enteritis. *Vet. Res.* **2015**, *46*, 98. [CrossRef]

41. He, Q.; Riley, R.T.; Sharma, R.P. Myriocin prevents fumonisin B1-induced sphingoid base accumulation in mice liver without ameliorating hepatotoxicity. *Food Chem. Toxicol.* **2005**, *43*, 969–979. [CrossRef]

42. Bhandari, N.; He, Q.; Sharma, R.P. Gender-related differences in subacute fumonisin B1 hepatotoxicity in BALB/c mice. *Toxicology* **2001**, *165*, 195–204. [CrossRef]

43. Sozmen, M.; Devrim, A.K.; Tunca, R.; Bayezit, M.; Dag, S.; Essiz, D. Protective effects of silymarin on fumonisin B$_1$-induced hepatotoxicity in mice. *J. Vet. Sci.* **2014**, *15*, 51–60. [CrossRef] [PubMed]

44. Hills, R.D., Jr.; Pontefract, B.A.; Mishcon, H.R.; Black, C.A.; Sutton, S.C.; Theberge, C.R. Gut Microbiome: Profound Implications for Diet and Disease. *Nutrients* **2019**, *11*, 1613. [CrossRef]

45. Lee, W.J.; Hase, K. Gut microbiota-generated metabolites in animal health and disease. *Nat. Chem. Biol.* **2014**, *10*, 416–424. [CrossRef]
46. Singh, M.P.; Kang, S.C. Endoplasmic reticulum stress-mediated autophagy activation attenuates fumonisin B1 induced hepatotoxicity in vitro and in vivo. *Food Chem. Toxicol.* **2017**, *110*, 371–382. [CrossRef] [PubMed]
47. Régnier, M.; Gourbeyre, P.; Pinton, P.; Napper, S.; Laffite, J.; Cossalter, A.M.; Bailly, J.D.; Lippi, Y.; Bertrand-Michel, J.; Bracarense, A.; et al. Identification of Signaling Pathways Targeted by the Food Contaminant FB1: Transcriptome and Kinome Analysis of Samples from Pig Liver and Intestine. *Mol. Nutr. Food Res.* **2017**, *61*, 1700433. [CrossRef] [PubMed]
48. Downes, J.; Dewhirst, F.E.; Tanner, A.C.R.; Wade, W.G. Description of Alloprevotella rava gen. nov., sp. nov., isolated from the human oral cavity, and reclassification of Prevotella tannerae Moore et al. 1994 as Alloprevotella tannerae gen. nov., comb. nov. *Int. J. Syst. Evol. Microbiol.* **2013**, *63*, 1214–1218. [CrossRef]
49. Li, Y.; Guo, Y.; Wen, Z.; Jiang, X.; Ma, X.; Han, X. Weaning Stress Perturbs Gut Microbiome and Its Metabolic Profile in Piglets. *Sci. Rep.* **2018**, *8*, 18068. [CrossRef] [PubMed]
50. Wan, X.; Li, T.; Liu, D.; Chen, Y.; Liu, Y.; Liu, B.; Zhang, H.; Zhao, C. Effect of Marine Microalga Chlorella pyrenoidosa Ethanol Extract on Lipid Metabolism and Gut Microbiota Composition in High-Fat Diet-Fed Rats. *Mar. Drugs* **2018**, *16*, 498. [CrossRef] [PubMed]
51. Alkadhi, S.; Kunde, D.; Cheluvappa, R.; Randall-Demllo, S.; Eri, R. The murine appendiceal microbiome is altered in spontaneous colitis and its pathological progression. *Gut Pathog.* **2014**, *6*, 25. [CrossRef]
52. Liang, X.; Bushman, F.D.; FitzGerald, G.A. Rhythmicity of the intestinal microbiota is regulated by gender and the host circadian clock. *Proc. Natl. Acad. Sci. USA* **2015**, *112*, 10479–10484. [CrossRef]
53. Zhang, W.; Jiao, L.; Liu, R.; Zhang, Y.; Ji, Q.; Zhang, H.; Gao, X.; Ma, Y.; Shi, H.N. The effect of exposure to high altitude and low oxygen on intestinal microbial communities in mice. *PLoS ONE* **2018**, *13*, e0203701. [CrossRef]
54. Piccolo, B.D.; Mercer, K.E.; Bhattacharyya, S.; Bowlin, A.K.; Saraf, M.K.; Pack, L.; Chintapalli, S.V.; Shankar, K.; Adams, S.H.; Badger, T.M.; et al. Early Postnatal Diets Affect the Bioregional Small Intestine Microbiome and Ileal Metabolome in Neonatal Pigs. *J. Nutr.* **2017**, *147*, 1499–1509. [CrossRef]
55. Drobac, D.; Tokodi, N.; Lujić, J.; Marinović, Z.; Subakov-Simić, G.; Dulić, T.; Važić, T.; Nybom, S.; Meriluoto, J.; Codd, G.A.; et al. Cyanobacteria and cyanotoxins in fishponds and their effects on fish tissue. *Harmful Algae* **2016**, *55*, 66–76. [CrossRef] [PubMed]
56. Puschner, B.; Moore, C. *Cyanobacteria*; W.B. Saunders: Saint Louis, MO, USA, 2013; pp. 533–540.
57. Clarke, G.; Stilling, R.M.; Kennedy, P.J.; Stanton, C.; Cryan, J.F.; Dinan, T.G. Minireview: Gut microbiota: The neglected endocrine organ. *Mol. Endocrinol.* **2014**, *28*, 1221–1238. [CrossRef] [PubMed]
58. Zinngrebe, J.; Montinaro, A.; Peltzer, N.; Walczak, H. Ubiquitin in the immune system. *EMBO Rep.* **2014**, *15*, 28–45. [CrossRef] [PubMed]
59. Huang, Z.; Rose, A.H.; Hoffmann, P.R. The role of selenium in inflammation and immunity: From molecular mechanisms to therapeutic opportunities. *Antioxid. Redox Signal.* **2012**, *16*, 705–743. [CrossRef] [PubMed]
60. Lennicke, C.; Rahn, J.; Kipp, A.P.; Dojčinović, B.P.; Müller, A.S.; Wessjohann, L.A.; Lichtenfels, R.; Seliger, B. Individual effects of different selenocompounds on the hepatic proteome and energy metabolism of mice. *Biochim. Biophys. Acta Gen. Subj.* **2017**, *1861*, 3323–3334. [CrossRef]
61. Bieberich, E. Synthesis, Processing, and Function of N-glycans in N-glycoproteins. *Adv. Neurobiol.* **2014**, *9*, 47–70. [CrossRef]
62. Yu, R.K.; Nakatani, Y.; Yanagisawa, M. The role of glycosphingolipid metabolism in the developing brain. *J. Lipid Res.* **2009**, *50*, S440–S445. [CrossRef] [PubMed]
63. Chow, J.C.; Young, D.W.; Golenbock, D.T.; Christ, W.J.; Gusovsky, F. Toll-like receptor-4 mediates lipopolysaccharide-induced signal transduction. *J. Biol. Chem.* **1999**, *274*, 10689–10692. [CrossRef] [PubMed]
64. Park, B.S.; Lee, J.O. Recognition of lipopolysaccharide pattern by TLR4 complexes. *Exp. Mol. Med.* **2013**, *45*, e66. [CrossRef] [PubMed]
65. Pugin, J.; Schürer-Maly, C.C.; Leturcq, D.; Moriarty, A.; Ulevitch, R.J.; Tobias, P.S. Lipopolysaccharide activation of human endothelial and epithelial cells is mediated by lipopolysaccharide-binding protein and soluble CD14. *Proc. Natl. Acad. Sci. USA* **1993**, *90*, 2744–2748. [CrossRef]
66. Joint FAO/WHO Expert Committee on Food Additives Eighty-Third Meeting. Available online: http://www.fao.org/3/bq821e/bq821e.pdf (accessed on 20 August 2021).
67. Guidance for Industry: Fumonisin Levels in Human Foods and Animal Feeds. Available online: https://www.federalregister.gov/documents/2001/11/09/01-28104/guidance-for-industry-fumonisin-levels-in-human-foods-and-animal-feeds-availability (accessed on 20 August 2021).
68. Lozupone, C.; Knight, R. UniFrac: A new phylogenetic method for comparing microbial communities. *Appl. Environ. Microbiol.* **2005**, *71*, 8228–8235. [CrossRef]
69. Caporaso, J.G.; Kuczynski, J.; Stombaugh, J.; Bittinger, K.; Bushman, F.D.; Costello, E.K.; Fierer, N.; Peña, A.G.; Goodrich, J.K.; Gordon, J.I.; et al. QIIME allows analysis of high-throughput community sequencing data. *Nat. Methods* **2010**, *7*, 335–336. [CrossRef] [PubMed]
70. Langille, M.G.; Zaneveld, J.; Caporaso, J.G.; McDonald, D.; Knights, D.; Reyes, J.A.; Clemente, J.C.; Burkepile, D.E.; Vega Thurber, R.L.; Knight, R.; et al. Predictive functional profiling of microbial communities using 16S rRNA marker gene sequences. *Nat. Biotechnol.* **2013**, *31*, 814–821. [CrossRef] [PubMed]

Article

Simultaneous Detection of Seven *Alternaria* Toxins in Mixed Fruit Puree by Ultra-High-Performance Liquid Chromatography-Tandem Mass Spectrometry Coupled with a Modified QuEChERS

Jiali Xing [1,2,†], Zigeng Zhang [3,†], Ruihang Zheng [1,*], Xiaorong Xu [1], Lingyan Mao [1], Jingping Lu [1], Jian Shen [1], Xianjun Dai [3,*] and Zhenfeng Yang [2]

[1] Ningbo Academy of Product and Food Quality Inspection (Ningbo Fibre Inspection Institute), Ningbo 315048, China; hellojiali77@gmail.com (J.X.); 1711091100@mail.nbu.edu.cn (X.X.); 1711091102@mail.nbu.edu.cn (L.M.); 1711091103@mail.nbu.edu.cn (J.L.); 1711091104@mail.nbu.edu.cn (J.S.)

[2] College of Biological and Environmental Sciences, Zhejiang Wanli University, Ningbo 315100, China; yangzf@zwli.edu.cn

[3] College of Life Sciences, China Jiliang University, Hangzhou 310018, China; p1809085214@cjlu.edu.cn

[*] Correspondence: 1711091101@mail.nbu.edu.cn (R.Z.); 05a0904027@cjlu.edu.cn (X.D.); Tel.: +86-574-89078647 (R.Z.); +86-574-89077478 (X.D.)

[†] These authors contributed equally to this work.

Citation: Zheng, R.; Zhang, Z.; Xing, J.; Xu, X.; Mao, L.; Lu, J.; Shen, J.; Dai, X.; Yang, Z. Simultaneous Detection of Seven *Alternaria* Toxins in Mixed Fruit Puree by Ultra-High-Performance Liquid Chromatography-Tandem Mass Spectrometry Coupled with a Modified QuEChERS. *Toxins* **2021**, *13*, 808. https://doi.org/10.3390/toxins13110808

Received: 18 October 2021
Accepted: 11 November 2021
Published: 16 November 2021

Publisher's Note: MDPI stays neutral with regard to jurisdictional claims in published maps and institutional affiliations.

Abstract: The presence of *Alternaria* toxins (ATs) in fruit purees may cause potential harm to the life and health of consumers. As time passes, ATs have become the key detection objects in this kind of food. Based on this, a novel and rapid method was established in this paper for the simultaneous detection of seven ATS (tenuazonic acid, alternariol, alternariol monomethyl ether, altenuene, tentoxin, altenusin, and altertoxin I) in mixed fruit purees using ultra-high performance liquid chromatography-tandem mass spectrometry. The sample was prepared using the modified QuEChERS (quick, easy, cheap, effective, rugged, and safe) method to complete the extraction and clean-up steps in one procedure. In this QuEChERS method, sample was extracted with water and acetonitrile (1.5% formic acid), then salted out with NaCl, separated on an ACQUITY UPLC BEH C_{18} with gradient elution by using acetonitrile and 0.1% formic acid aqueous as eluent, and detected by UPLC-MS/MS under positive (ESI^+) and negative (ESI^-) electrospray ionization and MRM models. Results showed that the seven ATs exhibited a good linearity in the concentration range of 0.5–200 ng/mL with $R^2 > 0.9925$, and the limits of detection (LODs) of the instrument were in the range of 0.18–0.53 μg/kg. The average recoveries ranged from 79.5% to 106.7%, with the relative standard deviations (RSDs) no more than 9.78% at spiked levels of 5, 10, and 20 μg/kg for seven ATs. The established method was applied to the determination and analysis of the seven ATs in 80 mixed fruit puree samples. The results showed that ATs were detected in 31 of the 80 samples, and the content of ATs ranged from 1.32 μg/kg to 54.89 μg/kg. Moreover, the content of TeA was the highest in the detected samples (23.32–54.89 μg/kg), while the detection rate of Ten (24/31 samples) was higher than the other ATs. Furthermore, the other five ATs had similar and lower levels of contamination. The method established in this paper is accurate, rapid, simple, sensitive, repeatable, and stable, and can be used for the practical determination of seven ATs in fruit puree or other similar samples. Moreover, this method could provide theory foundation for the establishment of limit standard of ATs and provide a reference for the development of similar detection standard methods in the future.

Keywords: *Alternaria* toxins; modified QuEChERS method; ultra-high-performance liquid chromatography-tandem mass spectrometry; mixed fruit puree

Key Contribution: An accurate and reliable UPLC-MS/MS method coupled with a modified QuEChERS method was developed for simultaneous determination of seven *Alternaria* toxins in mixed fruit puree for the first time.

1. Introduction

As a filamentous fungus, *Alternaria* strains are widely distributed in low-temperature and humid environments. *Alternaria* is one of the main microorganisms that cause the decay of fruits and vegetables and other agricultural products during transportation, processing, and storage [1]. As a secondary metabolite secreted by *Alternaria* strains, *Alternaria* toxins (*A*Ts) isolated from *Alternaria* fungi have reached at least 70 compounds. They are commonly divided into five groups according to their great structural divergence [2], namely, (I) dibenzopyranone derivatives of alternariol (AOH), alternariol monomethyl ether (AME), alternene (ALT) and altenusin (ALS) [3]; (II) the tetramic acid derivatives tenuazonic acid (TeA); (III) the perylene derivatives altertoxins (ATX-I, ATX-II, ATX-III); (IV) glycerin tricarboxylic ester compounds (AAL toxin), which can be divided into AAL-TA and AAL-TB; and (V) cyclic tetrapeptides, such as tentoxin (Ten). On the basis of the teratogenic, carcinogenicity, and embryonic toxicity by *A*Ts to human and animals, it is necessary to attach importance to the existence of *A*Ts in food [4].

*A*Ts have been found in various processed products, such as fruits and juices [5], vegetable-based products [6], wheat-based products [7], beer [8], and oil [9]. As a popular processed product, fruit puree is usually made without the addition of any preservatives, flavors, pigments, and other chemicals [10]. Different from other fruit products, fruit puree is another fruit form, while concentrated juice and fruit pulp are artificially manufactured, resulting in loss of nutrients and taste. Therefore, fruit puree is hygienic, nutritious, and healthy, which is suitable as a vitamin supplement for infants, children, and the elderly. Mixed fruit purees contain a certain amount of water, sugar, and other nutrients, which provides suitable conditions for *A*T formation during its processing and storage. The panel of the European Food Safety Authority on contaminants in the food chain evaluated the risks to public health related to *A*Ts in food in 2011 [11] and 2016 [12]. The results in both years showed that infants and young children had the most dietary exposure to *A*Ts. For its unique taste and balanced nutrition, mixed fruit puree has gradually become the mainstream market of infant fruit products [13]. Therefore, special attention must be given to addressing the potential *A*Ts pollution in mixed fruit purees.

The detection of *A*Ts is a necessary step in the safety evaluation of food products contaminated by *A*Ts. The instrumental techniques for measuring *A*Ts mainly include gas chromatography (GC) coupled to a mass spectrometry (MS) detector [14], and liquid chromatography (LC) coupled to an ultraviolet detector [15], a diode array detector [16], enzyme linked immunosorbent assay [17], thin layer chromatography, or a MS detector [18]. Considering that most *A*Ts are stable and unvolatile, GC and GC tandem MS are seldom used for detecting *A*Ts. Nevertheless, recent studies have demonstrated the advantages of the ultra-high-performance liquid chromatography tandem mass spectrometry (UPLC-MS/MS) technique in the determination of *A*Ts in sunflower oil [9], drinking water [19], and wolfberry [20] with high efficiency, precision, and sensitivity [21]. The purpose of the sample pretreatment procedure is usually to extract *A*Ts from food matrices prior to instrumental analysis. Given the complexity of food matrices, sample preparation strategies like solid phase extraction (SPE) or QuEChERS extraction are often required to reach a satisfactory sensitivity [8,22]. Given the complicated SPE operation and the low recovery of some toxins, such as AOH, SPE is unsuitable for the pretreatment of *A*Ts [23]. QuEChERS is a quick, easy, cheap, effective, rugged, and safe sample pretreatment technology based on dispersive SPE and has been successfully used in detecting *A*Ts [24]. The QuEChERS approach has already been applied to 25 mycotoxins in cereals [25] and different kinds of mycotoxins, including the *A*Ts in barley [18] and pomegranate [5]. To date, the QuEChERS method coupled with a UPLC-MS/MS method has successfully been applied to the determination of six *A*Ts (AOH, ATS, TeA, AME, ALT, and Ten) in mango [26,27]. However, there is no report on the occurrence of these *A*Ts in mixed fruit puree at present. Considering that simultaneous detection of *A*Ts in mixed fruit puree is very important for infant health, there is an urgent need to establish a QuEChERS method coupled with UPLC-MS/MS to detect the *A*Ts in mixed fruit puree.

A reliable and sensitive detection of *A*Ts was achieved by optimizing the water addition, dehydrating agent, salting out agent, extraction solvent, extraction method, and adsorbent for the QuEChERS procedure. The purpose of this research was to establish a modified QuEChERS method coupled with UPLC-MS/MS for simultaneously determining seven *A*Ts (TeA, AOH, AME, ALT, Ten, ALS, and ATX-I, based on the availability of the standard substance) in mixed fruit purees. Moreover, this method does not require extraction column purification or expensive equipment such as a gel permeation chromatograph, and the pretreatment process is simple. The established method was applied to determine the contents of seven *A*Ts in 80 mixed fruit puree samples.

2. Results and Discussion

2.1. Optimization of Water Addition

The rapid detection of pesticide residues in fruits and vegetables through UPLC-MS/MS coupled with QuEChERS has been intensively researched and has revealed that adding a certain amount of water to vegetables and fruits with low water content improves the extraction effects [28]. Considering the high sweetness and viscosity of mixed fruit purees, adding a certain amount of water can increase the recovery rate. In this study, the effects of water dosage (0, 1, 2, 3, 4 and 5 g) on the extraction efficiency were studied. The results (Figure 1) indicated that the best extraction effect, with a recovery rate between 85.1% and 96.4%, was achieved by adding 3 g of water into 5 g of mixed fruit puree. This may be due to the fact that acetonitrile could be better immersed in the sample to improve the extraction effect by adding water [29]. When the amount of water added was more than 3 g, the recoveries of seven *A*Ts was decreased. We deduced that the increased proportion of water would diluted the organic solvents used to extract *A*Ts, resulting in a decreased extraction performance.

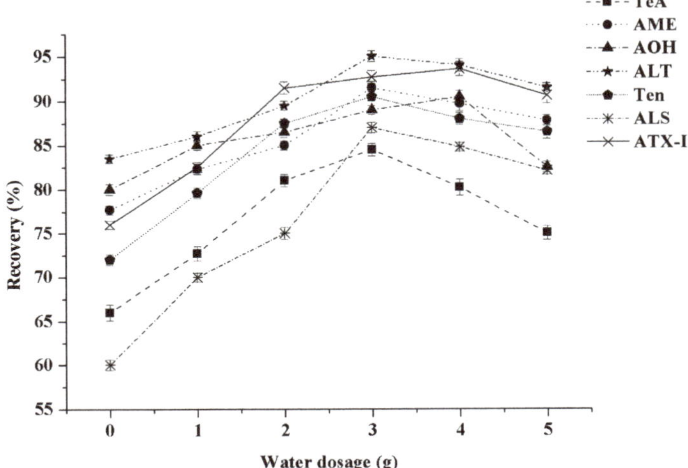

Figure 1. Effect of water content in mixed fruit mud on the recovery of seven kinds of *A*Ts (n = 3).

2.2. Optimization of Extraction Solvent

To minimize the interference of the co-extracted materials and improve the extraction efficiency of the seven *A*Ts, the extraction solvent was evaluated. Considering that TeA is highly acidic and more polar than other *A*Ts, it is easy to chelate with metals. Moreover, TeA usually exists in food in the form of salt. Therefore, adding a proper amount of acid to the extraction solvent is conducive to the TeA extraction [30]. In our research, the extraction effects of seven *A*Ts by pure acetonitrile; pure methanol; 1%, 1.5%, 2% FA acetonitrile; and 1%, 1.5%, 2% FA methanol solutions were compared. The results showed that the extraction solution was turbid and the recovery was only approximately 25% when pure methanol and 1%, 1.5%, 2% FA methanol solutions were used as extraction agents, which were much lower than those of the acetonitrile system, although the recoveries of seven *A*TS extracted by pure acetonitrile were low. The recoveries of seven target compounds were between 84% and 96% when the FA content reached 1.5% in acetonitrile, which were higher than those of other extraction solutions. Therefore, 1.5% FA acetonitrile solution was used as the extraction solution for seven *A*Ts. The recovery rates of the seven *A*Ts in the acetonitrile system are shown in Figure 2.

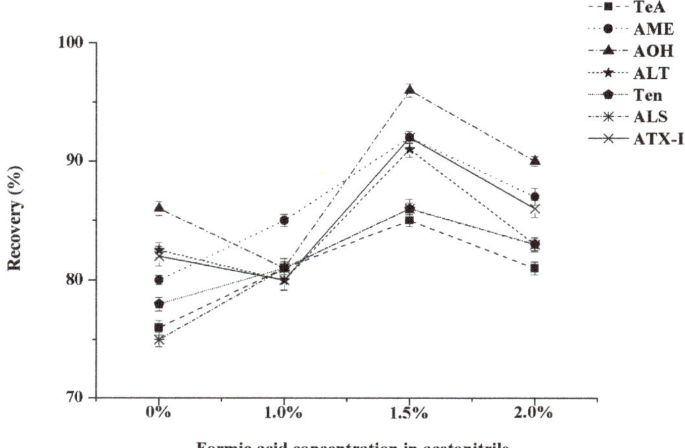

Figure 2. Effect of different FA concentrations on the recovery of seven kinds of *A*Ts in an acetonitrile system (n = 3).

Moreover, the dosage (5, 10, and 15 mL) of 1.5% FA acetonitrile solution on extraction efficiency was also investigated. The results showed that when adding 5 mL of the 1.5% FA acetonitrile solution, the recoveries of the seven *A*Ts were the highest, ranging from 84% to 95%. This result is consistent with the previous studies on the extraction of *A*Ts in fruits and vegetables by De et al. [6] and Dong et al. [31], who used the same extractant and dosage as the optimal extractant to extract *A*Ts. Therefore, 5 mL 1.5% FA acetonitrile solution was selected as the extraction agent in this study.

2.3. Optimization of Dehydrating Agent and Salting Out Agent

Anhydrous $MgSO_4$ is usually used to remove the moisture from the sample matrix in the QuEChERS method [32]. The effects of 0, 1, 2, 3, 4, and 5 g of anhydrous $MgSO_4$ on the recoveries of the seven *A*Ts were compared in this study. The results indicated that 2 g of anhydrous $MgSO_4$ was the optimal dosage of dehydrating agent for the six *A*Ts (Ten, AOH, AME, ALT, ALS, and ATX-I) in the mixed fruit puree samples, while the recovery of TeA was unsatisfactory. As shown in Figure 3, compared with the low recovery of TeA (15–32%) when the anhydrous $MgSO_4$ was added, the recovery of TeA was much higher (86%) without anhydrous $MgSO_4$. It may be due to the strong chelation of TeA on Mg^{2+}, which resulted in a decrease of recovery rate. Meanwhile, the recoveries of the other six *A*Ts had no significant difference whether the anhydrous $MgSO_4$ was added or not. Our results were similar to those obtained by Cheng et al. [33] and Chen et al. [34] on the detection of *A*Ts in red jujube and fruits. Thus, anhydrous $MgSO_4$ was not used in this study.

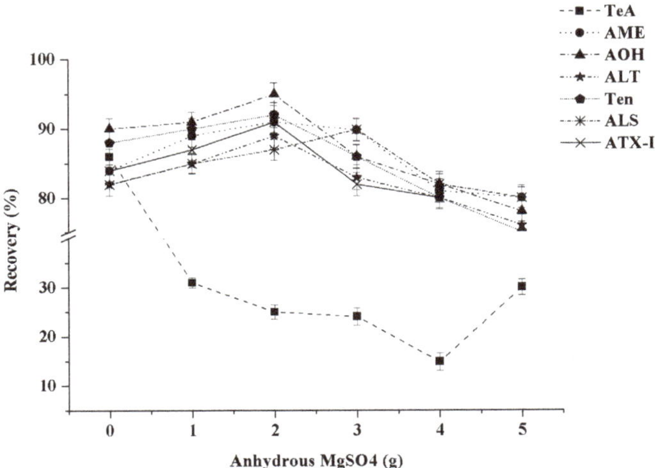

Figure 3. Effect of the dosages of anhydrous MgSO$_4$ on the recoveries of seven *A*Ts (n = 3).

With the addition of salting-out agent, the organic phase molecules in the extract will break the hydrogen bond with the water molecules due to the increase in ionic strength, and be salted out from the water, and the extraction efficiency can be greatly improved [35]. The salting out efficiencies of different NaCl dosages (0, 0.5, 1, 2, 3, 5 g) were evaluated in our study. As shown in Figure 4, as the dosage of NaCl increased, the recovery rates of the *A*Ts generally increased first and then decreased. Among them, when the dosage of NaCl was 2 g, the recovery rates of seven *A*Ts were the highest (85.5–96.8%). Therefore, 2 g of NaCl was selected as the salting-out agent in this study.

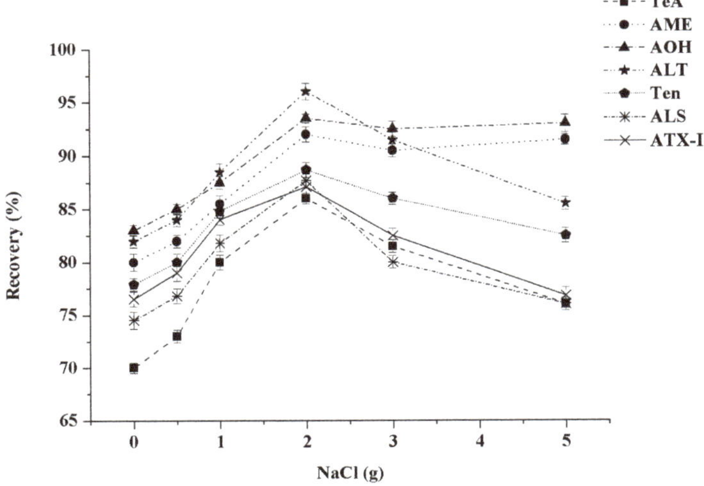

Figure 4. Effect of the dosages of NaCl on the recoveries of seven *A*Ts (n = 3).

2.4. Optimization of the QuEChERS Purification

The extraction solution showed a deep color after the mixed fruit puree samples were extracted with 1.5% FA in acetonitrile. This may be attributed to impurities, such as natural pigments, which were also extracted into the solution, so the extraction solutions must be purified further to reduce the influence of impurities [36]. QuEChERS purification techniques have been widely applied in the agricultural products and food detection fields [20]. Some adsorbents, such as octadecylsilyl (C18), primary secondary amine (PSA), and graphitized carbon black (GCB), are commonly employed in QuEChERS procedures. In this experiment, the purification effects of different adsorbents and whether to add adsorbent were studied. Since the GCB adsorbent has the same planar structure as the seven *AT*s, GCB could absorb *AT*s while absorbing impurities, so GCB was not considered as a purification adsorbent in this study [31]. Then, 2 mL of the upper extraction solvent was accurately transferred into 5 mL centrifuge tube pre-loaded with 50, 100, and 150 mg of C18 and PSA, respectively. The extraction solutions were vortexed and centrifuged. Subsequently, 500 µL of supernatant was mixed with 500 µL of primary water, which was filtered through a 0.22 µm organic filter membrane before detection by UPLC-MS/MS, and then the recoveries were calculated. The upper extractant detection without adsorbent was the same as the above operation.

The purification efficiencies for the seven *AT*s with the PSA adsorbent were higher than those obtained with C18 (Table 1). Moreover, the purification efficiencies were the highest when the amount of adsorbent was 100 mg. The effect without an adsorbent was similar to that with 100 mg of PSA. In addition, the precision and repeatability of the adsorbent were not ideal with the PSA adsorbent (Table 1). Jiang et al. [32] found that the addition of an adsorbent had no significant effect on the recovery rate of the *AT*s of citrus, so they did not choose the adsorbent. Besides, Guo et al. [37] found that the effect without any adsorbent was significantly higher than that with any other adsorbent in detecting the *AT*s of grapes. Finally, no adsorbent was added in the extraction process of *AT*s in mixed fruit puree.

Table 1. Purification efficiencies of different adsorbent types and amounts for the seven *AT*s.

Adsorbent	Recovery (%)						
	TeA	AME	AOH	ALT	Ten	ALS	ATX-I
0 mg	84.6 ± 1.2 [a]	92.1 ± 1.6 [a]	95.6 ± 2.1 [a]	90.3 ± 0.9 [c]	88.2 ± 1.7 [a]	86.5 ± 2.8 [b]	91.4 ± 1.7 [a]
50 mg C18	80.0 ± 3.8 [d]	86.9 ± 2.9 [g]	90.1 ± 3.6 [e]	87.7 ± 4.4 [f]	80.4 ± 4.6 [g]	82.6 ± 5.1 [e]	85.3 ± 3.8 [g]
100 mg C18	81.4 ± 3.3 [c]	88.9 ± 4.6 [e]	92.5 ± 2.9 [d]	90.3 ± 4.6 [c]	83.6 ± 3.1 [d]	84.3 ± 2.9 [d]	87.7 ± 3.9 [e]
150 mg C18	75.8 ± 4.0 [f]	87.2 ± 3.8 [f]	86.4 ± 3.1 [g]	88.6 ± 2.9 [e]	81.9 ± 2.2 [f]	78.9 ± 3.4 [g]	86.7 ± 2.2 [f]
50 mg PSA	80.1 ± 3.1 [d]	89.7 ± 3.6 [d]	93.2 ± 2.2 [c]	90.5 ± 2.8 [b]	84.4 ± 2.9 [c]	85.6 ± 3.1 [c]	89.1 ± 3.2 [c]
100 mg PSA	83.7 ± 3.9 [b]	91.6 ± 4.5 [b]	94.1 ± 5.2 [b]	91.2 ± 4.3 [a]	85.9 ± 4.6 [b]	87.7 ± 2.9 [a]	90.3 ± 5.6 [b]
150 mg PSA	77.5 ± 2.8 [e]	90.2 ± 5.4 [c]	87.6 ± 4.2 [f]	89.3 ± 3.9 [d]	83.4 ± 4.1 [e]	82.3 ± 3.3 [f]	88.6 ± 4.1 [d]

Note: Different letters in the same column represent significant differences ($p < 0.05$).

2.5. Optimization of the Extraction Method

In this experiment, the efficiency of different extraction methods was studied. They included vortex oscillation (300 r/min, 10 min), homogenization (12,000 r/min, 5 min), ultrasonic bath (40 °C, 20 min), and water bath oscillation (40 °C, 20 min) on the recovery of seven *AT*s were compared. As shown in Figure 5, when homogenous extraction was used the recoveries of ALT and TeA were 44% and 23%, respectively. We infer that puree sample stuck to the head of the homogenizer during the homogenization extraction process, resulting in excessive substrate loss and severely reducing the extraction effect of some *AT*s. As the same time, the extractant was not fully contacted with the sample located in the bottom of the centrifuge tube during ultrasonic extraction. Thus, the recoveries of TeA, AOH, and AME were all lower than 40% in ultrasonic extraction. Similarly, under the condition of water bath oscillation, the recoveries of TeA and ALS were also low, at 41% and 38%, respectively. Thus, we could deduce that moderate extractions like water bath oscillations, homogenization, and ultrasonic extraction were not suitable for the extraction of seven *AT*s. However, the recoveries of the seven *AT*s in vortex oscillation extraction were significantly higher than those of the other three extraction methods ($p < 0.05$), which exceeded 83%. This may be due to the fact that *A*Ts in the matrix are not easily destroyed during vortex oscillation.

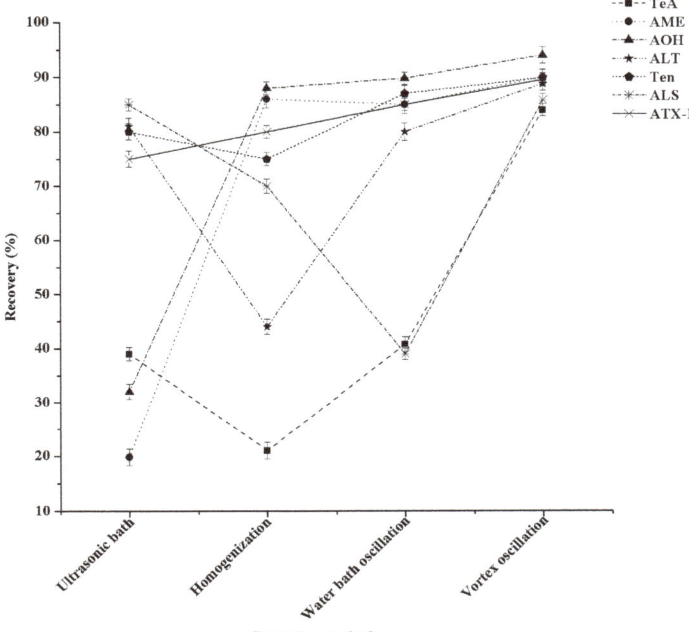

Figure 5. Effect of the four extraction methods on the recoveries of seven kinds of *A*Ts (n = 3).

In addition, the effects of vortex oscillation time (5, 10, and 20 min) on the recoveries of the seven *A*Ts were also compared. When the oscillation time was 10 min, the recoveries of seven *A*Ts were the highest (82.1–96.8%). Therefore, 10 min of vortex oscillation was selected as the extraction method in this study.

2.6. Optimization of Chromatography and Mass Spectrometry Conditions

Water-methanol and water-acetonitrile are commonly used in UPLC-MS/MS as the mobile phase [38]. Besides, the introduction of FA and ammonium formate can usually enhance the target response and improve the target peak [39]. In our previous study, FA acetonitrile was used as the extractant. To maintain consistency, this experiment focused on three mobile phase systems: water-acetonitrile, 0.1% FA aqueous solution-acetonitrile, and 0.1% FA with 5 mmol ammonium formate solution-acetonitrile. The results showed that the introduction of FA enhanced the response of the seven kinds of target *A*Ts, while the response of the target decreased after the introduction of ammonium formate, and trailing appeared in the peak type. Hence, 0.1% FA aqueous solution-acetonitrile was selected as the mobile phase system.

MRM ion mass spectra of the seven kinds of *A*Ts and the total MRM ion mass spectra of ATX-I (negative ions) and other six *A*Ts (positive ions) are shown in Figure 6, respectivly. The qualitative and quantitative ions of each toxin were determined through the continuous injection of the flow injection pump and then optimized by the mass spectrometry conditions (such as conic hole voltage, collision voltage, ion source temperature, desolvent gas temperature and flow, collision gas flow) to achieve the optimal ionization efficiency of each target substance. The samples were respectively scanned using ESI$^+$ and ESI$^-$ modes to find the parent ion with a high response value. The collision voltage was further changed and secondary mass spectrometry scanning was performed to find the daughter ions with strong signal and stability.

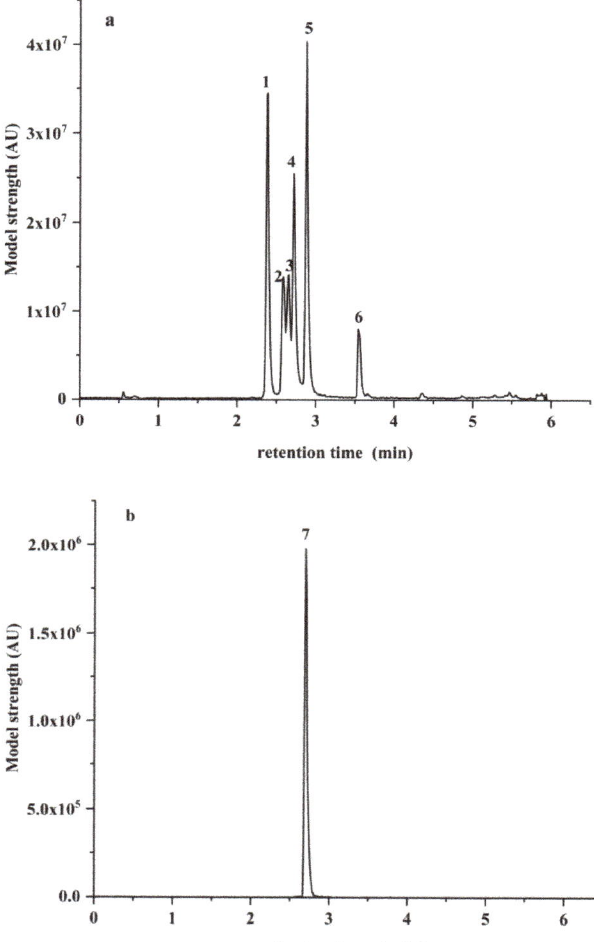

Figure 6. Mass spectrogram of seven *A*Ts under positive (**a**) and negative (**b**) electrospray ionization. Note: 1 for ALT; 2 for ALS; 3 for TeA; 4 for AOH; 5 for Ten; 6 for AME; 7 for ATX-I.

2.7. Method Validation

2.7.1. Matrix Effects (MEs)

As with ion enhancement or inhibition, MEs are caused by the influence of co-eluting compounds on the ionization efficiency of the electrospray interface in UPLC-MS/MS analysis [40]. Reportedly, MEs are common in *A*Ts analysis by UPLC-MS/MS [21]. In order to test whether the response value of the blank sample matrix to the target compound was enhanced or inhibited, seven kinds of *A*T mixed solutions were prepared with the solvent standard and matrix blank solution, respectively, at the concentration of 50 ng/mL, and the results were further compared.

As shown in Table 2, the ME values of Ten and ATX-I were between 80% and 100%, which indicated that the MEs of Ten and ATX-I could be ignored. TeA, AOH, and AME showed matrix suppression with ME values lower than 80%. On the contrary, ALT and ALS exhibited matrix enhancement with ME values higher than 120%. All of these indicated that some interfering substances still exist although the extractant solution was purified, which inhibited the analysis of the target analytes remained in the solution. This phenomenon was consistent with the results of other researches [41].

Table 2. Influence of dilution and small volume injection on the MEs.

Target Analyte	Matrix Effects before Dilution (Injection Volume: 10 μL)	Matrix Effects after Dilution (Injection Volume: 10 μL)	Matrix Effects after Dilution (Injection Volume: 3 μL)
TeA	69.7 ± 2.4 [a]	79.2 ± 1.4 [b]	85.9 ± 1.9 [c]
AME	78.6 ± 4.6 [a]	82.3 ± 3.2 [a]	88.3 ± 2.1 [b]
AOH	75.3 ± 3.1 [a]	87.1 ± 2.8 [b]	88.6 ± 1.7 [b]
ALT	123.3 ± 2.5 [c]	112.4 ± 3.2 [b]	98.7 ± 0.9 [a]
Ten	88.4 ± 1.8 [a]	90.3 ± 2.7 [ab]	92.2 ± 1.1 [b]
ALS	136.5 ± 3.4 [c]	120.5 ± 1.5 [b]	96.3 ± 1.8 [a]
ATX-I	90.1 ± 2.0 [a]	91.7 ± 1.3 [a]	93.1 ± 2.3 [b]

Note: Different lines in the same column represent significant differences ($p < 0.05$).

To compensate for the MEs, dilution (5 ng/mL) and a small volume injection (3 μL and 10 μL) were used to quantify the seven ATs in the mixed puree samples. The MEs of AME, Ten, and ATX-I had no significant change ($p > 0.05$) after dilution and injection with 10 μL. However, significantly different MEs ($p < 0.05$) were found for all the seven ATs after dilution and injection with 3 μL. Moreover, the MEs of all the seven ATs were between 80% and 100%. This indicated that both matrix suppression (TeA, AOH, and AME) and matrix enhancement (ALT and ALS) of seven ATs were resolved by dilution and injection with 3 μL (Table 3).

2.7.2. Linearity and Detectability of the Method

In the linearity studies, all the standard working solutions were determined under optimal chromatography and mass spectrometry conditions. Linear regression analysis was performed on a plot with concentration on the X-axis, and the peak area on the Y-axis. The results shown in Table 3 indicate that suitable linearities were obtained in the corresponding concentration range and the coefficients of determination (R^2 values) exceeded 0.990 for all seven ATs.

Table 3. Linear range, linear equation, R^2, and detection limit of seven kinds of ATs.

Component	Linear Range (ng/mL)	Linear Equation	R^2	LODs (μg/kg)	LOQs (μg/kg)
TeA	0.5–200	$y = 41232.3x - 3133.37$	0.9963	0.46	1.47
AME	0.5–200	$y = 2828.31x - 893.32$	0.9998	0.37	1.22
AOH	0.5–200	$y = 2503.73x - 1257.22$	0.9997	0.53	2.17
ALT	0.5–200	$y = 7573.01x + 248.023$	0.9996	0.22	0.77
Ten	0.5–200	$y = 16149.8x - 3371.39$	0.9998	0.18	0.56
ALS	0.5–200	$y = 1398.39x + 618.519$	0.9925	0.39	1.25
ATX-I	0.5–200	$y = 3136.11x - 1402.31$	0.9996	0.27	0.89

The LODs and LOQs of the method were calculated according to the validated experimental results. The results showed that the LODs of the seven ATs were in the range of 0.18–0.53 μg/kg, and the LOQs of the seven ATs were in the range of 0.56–2.17 μg/kg (Table 4).

2.7.3. Trueness and Precision of Standard Addition

The trueness and precision of the method were assessed for each toxin by determining the recoveries and the RSDs from the blank mixed fruit puree samples spiked at three different levels (5, 10, and 20 µg/kg). The average recoveries were in the range of 79.5–106.7%, and the RSDs were lower than 9.78% (Table 4). Thus, the trueness and precision of the seven *A*Ts in the mixed fruit purees are acceptable, satisfying the AOAC criteria [42].

Table 4. Trueness and precision of the optimized method (n = 3).

Component	Spiked (µg/kg)	Average Recovery (%)	RSD (%)
TeA	5	85.3	9.78
	10	88.2	8.65
	20	79.5	9.65
AME	5	93.0	8.85
	10	93.5	6.54
	20	106.7	5.63
AOH	5	87.2	5.36
	10	96.1	2.35
	20	102.8	7.21
ALT	5	85.6	6.08
	10	90.2	4.68
	20	98.6	6.31
Ten	5	90.3	3.67
	10	88.9	3.69
	20	101.5	5.48
ALS	5	86.0	4.56
	10	86.3	5.13
	20	83.2	5.48
ATX-I	5	91.1	3.43
	10	98.7	2.68
	20	96.5	5.45

2.7.4. Analysis of Fruit Puree Samples

This study established a method for simultaneously determining seven *A*Ts in mixed fruit purees by UPLC-MS/MS coupled with modified QuEChERS.

For the latter procedure, the modified QuEChERS method optimized the water addition, the extraction agent, the dehydrating agent, the salting out agent, the QuEChERS purification, and extraction method to make the pretreatment simpler and more effective. The optimized results showed that the recovery rates of the seven *A*Ts were the highest under the following conditions: 3 mL of primary water was added, 5 mL of 1.5% FA was used as the extraction agent and was extracted by vortex oscillation, no anhydrous $MgSO_4$ was used, 2 g of NaCl was used as the salting out agent, and no purifier was added. The established UPLC-MS/MS method can accurately, quickly, and reliably determine *A*Ts. The proposed method has satisfactory applicability and can be used in the risk monitoring of laboratories.

A total of 80 fruit puree samples for infants were determined by the established and validated method. The results showed that the seven *A*Ts were detected in 38.75% (31/80) of the mixed puree samples (Table 5). Besides, the content of TeA was the highest in the detected samples (23.32–54.89 µg/kg) while the detection rate of Ten (24/31 samples) was higher than the other *A*Ts. Furthermore, the other five *A*Ts had similar and lower levels of contamination. For instance, AOH and AME were detected in seven and five samples, which ranged from 3.75 µg/kg to 8.11 µg/kg and 2.28 µg/kg to 9.83 µg/kg, respectively. The samples of ALT, ALS, and ATX-I were detected in one, three, and two cases, respectively. Among them, the content of ALT was 2.66 µg/kg, the contents of ATX-I were 6.43 µg/kg and 7.54 µg/kg, and the contents of ALS were 4.11–15.48 µg/kg. This was an evidence for the contamination of multiple *A*Ts in the mixed fruit puree samples. In general, the concentrations of *A*Ts in mixed fruit puree were higher than those found in cereal-, vegetable-, and (or) fruit-based infant products [43], meaning that they could pose potential health risks to consumers. Thus, monitoring systems should be strictly enforced. In addition, the prevention and control strategies for the pre- and post-processing procedures should be improved.

Table 5. Detection results of the mixed fruit puree samples.

Samples	TeA (µg/kg)	AME (µg/kg)	AOH (µg/kg)	ALT (µg/kg)	Ten (µg/kg)	ALS (µg/kg)	ATX-I (µg/kg)
6	38.92	ND	ND	ND	ND	ND	ND
9	ND	ND	ND	ND	3.26	ND	ND
11	43.31	9.83	7.15	ND	ND	ND	ND
17	ND	ND	ND	ND	5.21	6.56	ND
20	ND	ND	ND	ND	4.73	ND	ND
24	ND	ND	ND	ND	2.11	ND	ND
27	47.96	ND	8.11	ND	4.39	ND	ND
28	ND	ND	ND	ND	5.51	ND	ND
31	ND	ND	ND	ND	2.66	ND	ND
32	52.68	ND	ND	ND	3.67	ND	7.54
35	ND	6.32	7.49	ND	1.66	ND	ND
38	ND	ND	ND	ND	6.32	ND	ND
39	ND	ND	ND	ND	ND	4.11	ND
41	38.99	ND	ND	ND	4.68	ND	ND
43	44.77	ND	ND	ND	8.37	ND	ND
44	ND	ND	4.17	ND	ND	ND	ND
45	ND	ND	ND	2.66	5.18	ND	ND
48	ND	2.28	ND	ND	4.89	ND	ND
50	54.89	ND	ND	ND	2.56	ND	ND
51	43.32	ND	ND	ND	1.69	ND	ND
52	ND	ND	ND	ND	4.67	ND	ND
55	ND	2.61	3.75	ND	4.33	ND	ND
59	ND	ND	ND	ND	5.68	ND	ND
62	34.44	ND	ND	ND	3.65	ND	ND
66	ND	ND	ND	ND	4.66	ND	ND
68	23.32	ND	ND	ND	2.37	ND	ND
70	36.98	ND	5.99	ND	ND	ND	ND
71	ND	ND	ND	ND	1.32	15.48	ND
74	45.67	ND	ND	ND	ND	ND	6.43
76	ND	3.92	4.21	ND	ND	ND	ND
79	33.29	ND	ND	ND	6.98	ND	ND

Note: ND for not detection.

3. Conclusions

In summary, a modified QuEChERS method coupled with a UPLC-MS/MS method was developed and validated for the analysis of seven *A*Ts in mixed fruit puree samples. Under the optimized chromatography and mass spectrometry conditions, mixed fruit puree samples were extracted with 1.5% FA in acetonitrile after adding 3 g water and salting out with 2 g NaCl, without dehydrating and purifying agents. This optimization not only simplifies the procedure, but also improves the recovery rates of the seven *A*Ts. This method had good selectivity, accuracy, and precision when using matrix-matched calibration curves for quantification. The LODs of the method ranged from 0.18 µg/kg to 0.53 µg/kg, and the LOQs were in the range of 0.56–2.17 µg/kg. This method was successfully applied in determining the seven *A*Ts in 80 mixed fruit puree samples. Among all the collected mixed fruit mud samples, 31 samples contained *A*Ts exceeding the levels of LODs, and the content of *A*Ts ranged from 1.32 µg/kg to 54.89 µg/kg. In general, the established method showed good performance in *A*Ts detection with high sensitivity and repeatability, and therefore could be applied to the routine monitoring of *A*Ts in mixed fruit puree. In addition, the newly developed method could effectively identify whether the mixed fruit puree was infected by toxigenic *Alternaria* to ensure the safety of mixed fruit puree and bring economic benefits for the development of the mixed fruit puree industry.

4. Materials and Methods

4.1. Sample Collection

Different brands of mixed puree samples (80 samples) were collected from different production bases and various supermarkets in Ningbo City, Zhejiang Province, China. The samples were sealed and stored at 4 °C for future use. Mixed fruit puree was mixed and matched by two or three fruits, such as apple, orange, blueberry, kiwi fruit, strawberry, banana, mango, lemon, peach, prune, coconut, pineapple, and blackcurrant.

4.2. Chemicals, Reagents, and Standards

Formic acid (FA), acetic acid, and acetonitrile (HPLC-grade) were purchased from Merck Co. (Darmstadt, Germany). Analytical reagent-grade anhydrous magnesium sulfate ($MgSO_4$), anhydrous sodium acetate (CH_3COONa), and sodium chloride (NaCl) were supplied by Sinopharm Chemical Reagent Co., Ltd. (Shanghai, China). Primary secondary amine (PSA) and octadecylsilane (C18), which were used as adsorbents, were all provided by ANPEL Laboratory Technologies (Shanghai) Inc.

Standards of TeA (CAS: 610-88-8), AOH (CAS: 641-38-3), AME (CAS: 26984-49-5), ALT (CAS: 29752-43-0), ALS (CAS: 31186-12-6), ATX-I (CAS: 56258-32-3), and Ten (CAS: 28540-82-1) were all acquired from Anpu Experimental Technology Co., Ltd. (Shanghai, China), and the purities of all the standards exceeded 98%. Each standard substance was prepared by dissolving 1 mg of the amorphous powder in 10 mL acetonitrile to obtain 100 μg/mL standard stock solutions and kept in a refrigerator at −20 °C. The seven individual standard stock solutions were diluted to prepare 1 μg/mL mixed standard solution, which was stored at −4 °C in amber glass vials under darkness before use.

4.3. Detection and Quantification Method

The UPLC-MS/MS system used for the separation and quantitation of the seven *AT*s consisted of a Waters ACQUITY TM UPLC and a Xevo TQ-S mass spectrometer (Waters Technology (Shanghai) Co., Ltd., Shanghai, China). The chromatographic separation was performed on a BEH C18 analytical column (50 mm × 2.1 mm, 1.7 μm, Waters Technology (Shanghai) Co., Ltd., Shanghai, China), and the column temperature was maintained at 40 °C. The flow rate was maintained at 0.4 mL/min, and the injection volume was 3 μL. The mobile phases were water (containing 0.1% FA, *v/v*) and acetonitrile. A linear gradient elution procedure was adopted for the separation of the seven *AT*s. The procedure was as follows: 0–5.0 min, 10–95% (acetonitrile phase); 5.0–7.0 min, 95% (acetonitrile phase); 7.0–7.5 min, 95–10% (acetonitrile phase); and 7.5–10.0 min, 10% (acetonitrile phase).

The mass sepectrometer used a Z-spray electrospray ionization (ESI) source. The ion source parameters were as follows: positive and negative ion switching scanning, capillary voltage of 1.08 kV, source temperature 150 °C, desolvation temperature 600 °C, desolvation gas flow 1000 L/h, and cone gas flow of 150 L/h. The cone voltage (CV), the parent ions, the collision energy (EC), and the fragment ions were optimized for each *AT* using the MassLynx InterlliStar software (Table 6). The seven *AT*s were analyzed in the multiple reaction monitoring (MRM) mode. Data acquisition and processing were accomplished using the MassLynx ™ 4.2 software.

Table 6. MS parameters.

Component	Ionization Mode	Parent (*m/z*)	Daughter (*m/z*)	Dwell Time (s)	Cone Voltage (V)	Collision Voltage (V)
Ten	ESI⁺	415.4	199.2 * 171.2	0.012	25	13 18
AME	ESI⁺	273.2	258.2 128.1 *	0.012	25	25 40
AOH	ESI⁺	259.2	213.2 185.1 *	0.012	25	25 30
TeA	ESI⁺	198.2	125.1 * 153.1	0.012	25	15 12
ALT	ESI⁺	293.2	257.2 * 275.4	0.012	25	12 8
ALS	ESI⁺	291.2	255.2 199.2 *	0.012	25	18 30
ATX-I	ESI⁻	351.3	315.25 * 333.3	0.0.12	25	8 10

Note: * is quantitative ion.

4.4. Sample Pretreatment Method

Sample pretreatment is a key step in sample analysis, as it will affect the accuracy. It includes sample dilution, sample extraction, and sample purification. First, the dosage of the dilution solvent was optimized and the best dosage of dilution solvent was selected by comparing the effects of different water dosages on *A*Ts. Second, the extraction solvent was optimized. With acetonitrile, 1% FA in acetonitrile, 1.5% FA in acetonitrile, 2.0% FA in acetonitrile, methanol, 1% FA in methanol, 1.5% FA in methanol, and 2.0% FA in methanol as the extraction solvents, the best extraction solvent was selected by evaluating the extraction efficiencies with different proportions of these extraction solvents. The effect of the addition of extraction solvent on the extraction efficiency was also evaluated. Finally, the purification process was optimized. With GCB, PSA, and C18 as the adsorbents, the three levels were evaluated. The best adsorbent type and amount were selected by comparing the recoveries obtained with different types and amounts of adsorbent.

In total, 5 g (ME204E, Shanghai Mettler Toledo Instrument Co., Ltd., Shanghai, China) of mixed fruit puree was weighed into a 50 mL plastic centrifuge tube. Then, 3 mL of water, 5 mL of 1.5% FA in acetonitrile, and 2 g of NaCl were added sequentially into the tube. The mixture was vortexed for 10 min (Vortex 3, Guangzhou Yike Laboratory Technology Co., Ltd., ShenZhen, China) and then centrifuged for 5 min at 9500 r/min (TGL-20M, Luxiangyi Centrifuge Instrument Co., Ltd., Shanghai, China). Subsequently, 500 μL of supernatant and 500 μL of water were mixed with the vortex (Vortex 3, Guangzhou Yike Laboratory Technology Co., Ltd., Guangdong, China) and filtered through a 0.22 μm organic filter membrane. Finally, the supernatant was determined by UPLC-MS/MS.

4.5. Method Validation

Exhaustive validation of this newly developed methodology was carried out in terms of the matrix effects (MEs), selectivity, linearity, accuracy (recovery), precision (relative standard deviation).

The MEs were assessed by comparing the peak areas of the mixed matrix standard with those of the mixed solvent standard. The values of the MEs were split into three groups (80–120%, higher than 120% and lower than 80%) based on the determined *A*Ts values. The ME values between 80% and 120% were classified as low MEs, which can be ignored. When the ME values exceeded 120%, they were deemed as matrix enhancements. Meanwhile, the ME values lower than 80% could be classified as matrix suppression. The MEs could be calculated by the following formula [27]:

$$ME(\%) = \frac{A_2}{A_1} \times 100\%$$

In the formula, A_1 is the average peak area of the toxin standard in pure solvent (initial mobile phase) at a specific concentration, and A_2 is the average peak area of the toxin standard at the same concentration in the matrix blank solution.

To assess the linearity of the calibration curves, a mixed standard solution of seven *A*Ts was diluted into nine different concentrations (0.5, 1.0, 2.0, 5.0, 10.0, 20.0, 50.0, 100.0, and 200.0 μg/L) using blank mixed fruit puree matrix. The linear equations of the calibration curves were obtained by plotting the concentrations of the seven *A*Ts and the corresponding peak areas, and the correlation coefficients (R^2 values) were calculated. The limits of detection (LODs) and limits of quantification (LOQs) of the seven *A*Ts were determined by serially diluting a mixed standard solution with blank mixed fruit puree matrix solution. The LODs were determined when the signal to noise ratio (S/N) was higher than or equal to 3, and the LOQs were taken when the S/N was higher than or equal to 10.

To evaluate the trueness and precision of the method, mixed standard solution of seven *A*Ts was added to blank mixed fruit puree samples at three different concentrations (5, 10 and 20 μg/kg), and the spiked samples were determined under the optimized pretreatment and analysis conditions. The spiked samples were determined 3 times and the relative standard deviations (RSDs) of the seven *A*Ts were calculated.

Author Contributions: J.X. and Z.Z. conceived and designed the experiments; R.Z. and X.X. performed the experiments; L.M. and J.L. analyzed the data; J.S. and X.D. and Z.Y. wrote the paper. All authors have read and agreed to the published version of the manuscript.

Funding: This work was supported by the Science and Technology Plan Program of State Administration for Market Regulation (2019MK080), the Project of Public Service Technology Application Analysis and Test of Zhejiang Science and Technology Department (LGC20C200003), the Project of Natural Science Foundation of Ningbo (No. 2019A610438, 2019A610437, 202003N4196), Fan-3315 Innovation Team of Ningbo (No. 2018B-18-C), the Public Welfare Research Project of Ningbo(2021S193), the Key Scientific Research Project of Ningbo (2021ZDYF020179)and a project funded by the High-Tech Elite Innovation Team of Ningbo (Yonggaoke [2018] No.63).

Institutional Review Board Statement: Not applicable.

Informed Consent Statement: Not applicable.

Data Availability Statement: Not applicable.

Acknowledgments: Thanks to Ningbo Academy of Product and Food Quality Inspection (Ningbo Fibre Inspection Institute) for providing experimental sites and instruments for this study.

Conflicts of Interest: The authors declare no conflict of interest.

References

1. Chen, Y.M. Determination of Alternaria Toxins in Fruits by Liquid Chromatography. Master's Thesis, Shanxi Normal University, Xi'an, China, 2012.
2. King, A.D.; Schade, J.E. Alternaria Toxins and Their Importance in Food. *J. Food Prot.* **1984**, *47*, 886–901. [CrossRef] [PubMed]
3. Puntscher, H.; Kutt, M.L.; Skrinjar, P.; Mikula, H.; Podlech, J.; Frohlich, J.; Marko, D.; Warth, B. Tracking emerging mycotoxins in food: Development of an LC-MS/MS method for free and modified Alternaria toxins. *Anal. Bioanal. Chem.* **2018**, *410*, 4481–4494. [CrossRef] [PubMed]
4. Puntscher, H.; Cobankovic, I.; Marko, D.; Warth, B. Quantitation of free and modified Alternaria mycotoxins in European food products by LC-MS/MS. *Food Control* **2019**, *102*, 157–165. [CrossRef]
5. Myresiotis, C.K.; Testempasis, S.; Vryzas, Z.; Karaoglanidis, G.S.; Papadopoulou-Mourkidou, E. Determination of mycotoxins in pomegranate fruits and juices using a QuEChERS-based method. *Food Chem.* **2015**, *182*, 81–88. [CrossRef]
6. De Berardis, S.; De Paola, E.L.; Montevecchi, G.; Garbini, D.; Masino, F.; Antonelli, A.; Melucci, D. Determination of four Alternaria alternata mycotoxins by QuEChERS approach coupled with liquid chromatography-tandem mass spectrometry in tomato-based and fruit-based products. *Food Res. Int.* **2018**, *106*, 677–685. [CrossRef] [PubMed]
7. Zhou, J.; Xu, J.J.; Cai, Z.X.; Huang, B.F.; Jin, M.C.; Ren, Y.P. Simultaneous determination of five Alternaria toxins in cereals using QuEChERS-based methodology. *J. Chromatogr. B Anal. Technol. Biomed. Life Sci.* **2017**, *1068*, 15–23. [CrossRef]
8. Prelle, A.; Spadaro, D.; Garibaldi, A.; Gullino, M.L. A new method for detection of five alternaria toxins in food matrices based on LC–APCI-MS. *Food Chem.* **2013**, *140*, 161–167. [CrossRef]

9. Tölgyesi, Á.; Kozma, L.; Sharma, V.K.; Kuo, P.-C. Determination of Alternaria Toxins in Sunflower Oil by Liquid Chromatography Isotope Dilution Tandem Mass Spectrometry. *Molecules* **2020**, *25*, 1685. [CrossRef]
10. Wang, Z.Z. Processing of four kinds of jam food. *Agric. Consult.* **2016**, *7*, 54.
11. Rose, M.D. Scientific Opinion on the risks for animal and public health related to the presence of Alternaria toxins in feed and food. *EFSA J.* **2011**, *9*, 2407.
12. Arcella, D.; Eskola, M.; Gómez Ruiz, J.A. Dietary exposure assessment to Alternaria toxins in the European population. *EFSA J.* **2016**, *14*, 4654.
13. Tanaka, F.; Shoji, Y.; Okazaki, K.; Miyazawa, T. Sensory Attributes of Apple Products Enhanced with Reconstituted Ethyl Esters that are Aroma Components of Watercored Apples. *J. Jpn. Soc. Food Sci. Technol.* **2017**, *64*, 34–37. [CrossRef]
14. Scott, P.M.; Weber, D.; Kanhere, S.R. Gas chromatography-mass spectrometry of Alternaria mycotoxins. *J. Chromatogr. A* **1997**, *765*, 255–263. [CrossRef]
15. Delgado, T.; Gómez-Cordovés, C.; Scott, P.M. Determination of alternariol and alternariol methyl ether in apple juice using solid-phase extraction and high-performance liquid chromatography. *J. Chromatogr. A* **1996**, *731*, 109–114. [CrossRef]
16. Michele, S.; Annalisa, D.G.; Carolina, V.; Angelo, V.; van den Bulk, R. Liquid chromatographic determination of Alternaria toxins in carrots. *J. AOAC Int.* **2004**, *87*, 101–106.
17. Gross, M.; Curtui, V.; Ackermann, Y.; Latif, H.; Usleber, E. Enzyme Immunoassay for Tenuazonic Acid in Apple and Tomato Products. *J. Agric. Food Chem.* **2011**, *59*, 12317–12322. [CrossRef]
18. Rubert, J.; Dzuman, Z.; Vaclavikova, M.; Zachariasova, M.; Soler, C.; Hajslova, J. Analysis of mycotoxins in barley using ultra high liquid chromatography high resolution mass spectrometry: Comparison of efficiency and efficacy of different extraction procedures. *Talanta* **2012**, *99*, 712–719. [CrossRef]
19. Zhang, Y.T.; Li, H.; Zhang, J.; Shao, B. Determination of Alternaria toxins in drinking water by ultra-performance liquid chromatography tandem mass spectrometry. *Environ. Sci. Pollut. Res. Int.* **2019**, *26*, 22485–22493. [CrossRef]
20. Xing, L.J.; Zou, L.J.; Luo, R.F.; Wang, Y. Determination of five Alternaria toxins in wolfberry using modified QuEChERS and ultra-high performance liquid chromatography-tandem mass spectrometry. *Food Chem.* **2020**, *311*, 1–27. [CrossRef]
21. Jeroen, W.; Hannes, M.; Michael, R.; Stefan, A.; Njumbe, E.E.; Diana, D.M.J.; Anita, V.L.; Lynn, V.; Sarah, D.S. Development and validation of an ultra-high-performance liquid chromatography tandem mass spectrometric method for the simultaneous determination of free and conjugated Alternaria toxins in cereal-based foodstuffs. *J. Chromatogr. A* **2014**, *1372*, 91–101.
22. Noser, J.; Schneider, P.; Rother, M.; Schmutz, H. Determination of six Alternaria toxins with UPLC-MS/MS and their occurrence in tomatoes and tomato products from the Swiss market. *Mycotoxin Res.* **2011**, *27*, 265–271. [CrossRef]
23. Zhou, Y.B.; Li, L.; Wu, Y.T.; Lin, Y.; Zhang, Q.; Bi, S.; Liu, W.Z.; Liu, L.Y. QuEChERS purification—ULTRA high performance liquid chromatography—tandem mass spectrometry for the determination of 5 kinds of Alternaria toxin in tomato. *Phys. Chem. Exam.* **2019**, *55*, 1036–1041.
24. Dong, L.N.; Chen, J.B.; Liu, J.; Zhao, M.M.; Ding, H.; Wang, X.H.; Dingjin, H.U.; Zhou, Y.X. Advanced in Application of QuEChERS for Multi-Mycotoxins Analysis in Food. *Acta Polym. Sin.* **2014**, *116*, 121–127.
25. Sun, J.; Li, W.X.; Zhang, Y.; Hu, X.X.; Wu, L.; Wang, B.J. QuEChERS Purification Combined with Ultrahigh-Performance Liquid Chromatography Tandem Mass Spectrometry for Simultaneous Quantification of 25 Mycotoxins in Cereals. *Toxins* **2016**, *8*, 375. [CrossRef] [PubMed]
26. Zhang, Z.G.; Zhang, A.Z.; Xing, J.L.; Cheng, H.; Zhang, S.F.; Zheng, R.H.; Dai, X.J. Determination residues of six alternaria toxins in mango by HPLC—MS. *Sci. Technol. Food Ind.* **2020**, *41*, 199–205, 211.
27. Stachniuk, A.; Fornal, E. Liquid Chromatography-Mass Spectrometry in the Analysis of Pesticide Residues in Food. *Food Anal. Methods* **2016**, *9*, 1654–1665. [CrossRef]
28. Mandal, S.; Poi, R.; Ansary, I.; Hazra, D.K.; Bhattacharyya, S.; Karmaker, R. Validation of a modified QuEChERS method to determine multiclass multipesticide residues in apple, banana and guava using GC–MS and LC–MS/MS and its application in real sample analysis. *SN Appl. Sci.* **2020**, *2*, 188. [CrossRef]
29. Shen, H.H.; Li, X.Y.; Su, M.; Hei, W.; He, T.T.; Wang, C. Rapid determination of 29 pesticide residues in Aksu jujube by QuEChERS-ultra performance liquid chromatography-tandem mass spectrometry. *J. Food Saf. Qual.* **2019**, *10*, 8410–8417.
30. Scott, P.M. Analysis of agricultural commodities and foods for Alternaria mycotoxins. *J. AOAC Int.* **2001**, *84*, 1809–1817. [CrossRef]
31. Dong, H.; Xian, Y.P.; Xiao, K.J.; Wu, Y.L.; Zhu, L.; He, J.P. Development and comparison of single-step solid phase extraction and QuEChERS clean-up for the analysis of 7 mycotoxins in fruits and vegetables during storage by UHPLC-MS/MS. *Food Chem.* **2019**, *274*, 471–479. [CrossRef]
32. Jiang, L.Y.; Zhao, Q.Y.; Gong, L.; Liu, Y.Y.; Zhang, Y.H.; Ma, L.; Jiao, B.N. Rapid determination of five Alternaria toxins species in citrus by SUPER-high performance liquid chromatography (HPLC) tandem mass spectrometry. *Chin. J. Anal. Chem.* **2015**, *43*, 1851–1858.
33. Cheng, J.X.; Li, Y.M.; Yang, Y.X.; Yang, Q.; Yuan, D.; Yang, J.; Wang, Y.C.; Zhang, R.L.; Du, Z.X. Determination and Risk Assessment of Four Alternaria Mycotoxins in Jujube. *J. Instrum. Anal.* **2018**, *37*, 1334–1338.
34. Kamikata, K.; Vicente, E.; Arisseto-Bragotto, A.P.; de Oliveira Miguel, A.M.R.; Milani, R.F.; Tfouni, S.A.V. Occurrence of 3-MCPD, 2-MCPD and glycidyl esters in extra virgin olive oils, olive oils and oil blends and correlation with identity and quality parameters. *Food Control* **2019**, *95*, 135–141. [CrossRef]

35. Hemmati, M.; Tejada-Casado, C.; Lara, F.J.; García-Campaña, A.M.; Rajabi, M.; del Olmo-Iruela, M. Monitoring of cyanotoxins in water from hypersaline microalgae colonies by ultra high performance liquid chromatography with diode array and tandem mass spectrometry detection following salting-out liquid-liquid extraction. *J. Chromatogr. A* **2019**, *1608*, 1–31. [CrossRef] [PubMed]
36. González-Jartín, J.M.; Alfonso, A.; Rodríguez, I.; Sainz, M.J.; Vieytes, M.R.; Botana, L.M. A QuEChERS based extraction procedure coupled to UPLC-MS/MS detection for mycotoxins analysis in beer. *Food Chem.* **2019**, *275*, 703–710. [CrossRef]
37. Guo, W.; Fan, K.; Nie, D.; Meng, J.; Huang, Q.; Yang, J.; Shen, Y.; Tangni, E.K.; Zhao, Z.; Wu, Y.; et al. Development of a QuEChERS-Based UHPLC-MS/MS Method for Simultaneous Determination of Six Alternaria Toxins in Grapes. *Toxins* **2019**, *11*, 87. [CrossRef]
38. Wang, C.C. Determination of Pesticide Residues in Fruits and Vegetables by Liquid Chromatography-Tandem Mass Spectrometry. Master's Thesis, Shandong University, Jinan, China, 2017.
39. Wang, C.; Fan, Y.; He, W.; Hu, D.; Wu, A.; Wu, W. Development and Application of a QuEChERS-Based Liquid Chromatography Tandem Mass Spectrometry Method to Quantitate Multi-Component Alternaria Toxins in Jujube. *Toxins* **2018**, *10*, 382. [CrossRef]
40. Wu, C.J.; Wu, L.; Li, H.; Yu, S.J. Determination of 4(5)-methylimidazole in foods and beverages by modified QuEChERS extraction and liquid chromatography-tandem mass spectrometry analysis. *Food Chem.* **2019**, *280*, 278–285. [CrossRef] [PubMed]
41. Wang, M.; Jiang, N.; Xian, H.; Wei, D.; Shi, L.; Feng, X. A single-step solid phase extraction for the simultaneous determination of 8 mycotoxins in fruits by ultra-high performance liquid chromatography tandem mass spectrometry. *J. Chromatogr. A* **2016**, *1429*, 22–29. [CrossRef]
42. De Smedt, J.M. AOAC validation of qualitative and quantitative methods for microbiology in foods. Association of Official Agricultural Chemists. *Int. J. Food Microbiol.* **1998**, *45*, 25–28. [CrossRef]
43. Marina, G.; Stefan, A.; Klara, G.; Fooladi, M.A.; Elisabeth, B.; Roland, K.; Michael, R. Quantitation of Six Alternaria Toxins in Infant Foods Applying Stable Isotope Labeled Standards. *Front. Microbiol.* **2019**, *10*, 1–14.

MDPI
St. Alban-Anlage 66
4052 Basel
Switzerland
Tel. +41 61 683 77 34
Fax +41 61 302 89 18
www.mdpi.com

Toxins Editorial Office
E-mail: toxins@mdpi.com
www.mdpi.com/journal/toxins